T0210746

Texts in Computer Science

Series Editors

David Gries, Department of Computer Science, Cornell University, Ithaca, NY, USA

Orit Hazzan ⓘ, Faculty of Education in Technology and Science, Technion—Israel Institute of Technology, Haifa, Israel

Titles in this series now included in the Thomson Reuters Book Citation Index!

'Texts in Computer Science' (TCS) delivers high-quality instructional content for undergraduates and graduates in all areas of computing and information science, with a strong emphasis on core foundational and theoretical material but inclusive of some prominent applications-related content. TCS books should be reasonably self-contained and aim to provide students with modern and clear accounts of topics ranging across the computing curriculum. As a result, the books are ideal for semester courses or for individual self-study in cases where people need to expand their knowledge. All texts are authored by established experts in their fields, reviewed internally and by the series editors, and provide numerous examples, problems, and other pedagogical tools; many contain fully worked solutions.

The TCS series is comprised of high-quality, self-contained books that have broad and comprehensive coverage and are generally in hardback format and sometimes contain color. For undergraduate textbooks that are likely to be more brief and modular in their approach, require only black and white, and are under 275 pages, Springer offers the flexibly designed Undergraduate Topics in Computer Science series, to which we refer potential authors.

Tomas Hrycej • Bernhard Bermeitinger •
Matthias Cetto • Siegfried Handschuh

Mathematical Foundations of Data Science

Tomas Hrycej
Institute of Computer Science
University of St. Gallen
St. Gallen, Switzerland

Bernhard Bermeitinger
Institute of Computer Science
University of St. Gallen
St. Gallen, Switzerland

Matthias Cetto
Institute of Computer Science
University of St. Gallen
St. Gallen, Switzerland

Siegfried Handschuh ⓘ
Institute of Computer Science
University of St. Gallen
St. Gallen, Switzerland

ISSN 1868-0941   ISSN 1868-095X (electronic)
Texts in Computer Science
ISBN 978-3-031-19076-6   ISBN 978-3-031-19074-2 (eBook)
https://doi.org/10.1007/978-3-031-19074-2

This Springer imprint is published by the registered company Springer Nature Switzerland AG
The registered company address is: Gewerbestrasse 11, 6330 Cham, Switzerland

Preface

Data Science is a rapidly expanding field with increasing relevance. There are correspondingly numerous textbooks about the topic. They usually focus on various Data Science methods. In a growing field, there is a danger that the number of methods grows, too, in a pace that it is difficult to compare their specific merits and application focus.

To cope with this *method avalanche*, the user is left alone with the judgment about the method selection. He or she can be helped only if some basic principles such as fitting model to data, generalization, and abilities of numerical algorithms are thoroughly explained, independently from the methodical approach. Unfortunately, these principles are hardly covered in the textbook variety. This book would like to close this gap.

For Whom Is This Book Written?

This book is appropriate for advanced undergraduate or master's students in computer science, Artificial Intelligence, statistics or related quantitative subjects, as well as people from other disciplines who want to solve Data Science tasks. Elements of this book can be used earlier, e.g., in introductory courses for Data Science, engineering, and science students who have the required mathematical background.

We developed this book to support a semester course in Data Science, which is the first course in our Data Science specialization in computer science. To give you an example of how we use this book in our own lectures, our Data Science course consists of two parts:

- In the first part, a general framework for solving Data Science tasks is described, with a focus on facts that can be supported by mathematical and statistical arguments. This part is covered by this book.
- In the second part of the course, concrete methods from multivariate statistics and machine learning are introduced. For this part, many well-known Springer textbooks are available (e.g., those by Hastie and Tibshirani or Bishop), which are used to accompany this part of the course. We did not intend to duplicate this voluminous work in our book.

Besides students as the intended audience, we also see a benefit for researchers in the field who want to gain a proper understanding of the mathematical foundations instead of sole computing experience as well as practitioners who will get mathematical exposure directed to make clear the causalities.

What Makes This Book Different?

This book encompasses the formulation of typical tasks as input/output mappings, conditions for successful determination of model parameters with good generalization properties as well as convergence properties of basic classes of numerical algorithms used for parameter fitting.

In detail, this book focuses on topics such as

- generic type of Data Science task and the conditions for its solvability;
- trade-off between model size and volume of data available for its identification and its consequences for model parametrization (frequently referred to as *learning*);
- conditions to be satisfied for good performance of the model on novel data, i.e., *generalization*; and
- conditions under which numerical algorithms used in Data Science operate and what performance can be expected from them.

These are fundamental and omnipresent problems of Data Science. They are decisive for the success of the application, more than a detailed selection of a computing method. These questions are scarcely, or not at all, treated in other Data Science and Machine Learning textbooks. Students and many data engineers and researchers are frequently not aware of these conditions and, neglecting them, produce suboptimal solutions.

In this book, we did not focus on Data Science technology and methodology except where it is necessary to explain general principles, because we felt that this was mostly covered in existing books.

In summary, this textbook is an important addition to all existing Data Science courses.

Comprehension Checks

In all chapters, important theses are summarized in their own paragraphs. All chapters have comprehension checks for the students.

Acknowledgments

During the writing of this book, we have greatly benefited from students taking our course and providing feedback on earlier drafts of the book. We would like to explicitly mention the help of Jonas Herrmann for thorough reading of the manuscript. He gave us many helpful hints for making the explanations comprehensible, in particular from a student's viewpoint. Further, we want to thank Wayne Wheeler and Sriram Srinivas from Springer for their support and their patience with us in finishing the book.

Finally, we would like to thank our families for their love and support.

St. Gallen, Switzerland   Tomas Hrycej
September 2022   Bernhard Bermeitinger
                                                                              Matthias Cetto
                                                                         Siegfried Handschuh

Contents

Acronyms

AI	Artificial Intelligence
ARMA	Autoregressive Moving Average
BERT	Bidirectional Encoder Representations from Transformers
CNN	Convolutional Neural Network
CV	Computer Vision
DL	Deep Learning
DS	Data Science
FIR	Finite Impulse Response
GRU	Gated Recurrent Unit
IIR	Infinite Impulse Response
ILSVRC	ImageNet Large Scale Visual Recognition Challenge
LSTM	Long Short-Term Memory Neural Network
MIMO	Multiple Input/Multiple Output
MSE	Mean Square Error
NLP	Natural Language Processing
OOV	Out-of-Vocabulary
PCA	Principal Component Analysis
ReLU	Rectified linear units
ResNet	Residual Neural Network
RNN	Recurrent Neural Network
SGD	Stochastic Gradient Descent
SISO	Single Input/Single Output
SVD	Singular value decomposition
SVM	Support vector machine

Data Science and Its Tasks

As the name *Data Science* (DS) suggests, it is a scientific field concerned with data. However, this definition would encompass the whole of information technology. This is not the intention behind delimiting the Data Science. Rather, the focus is on *extracting useful information from data*.

In the last decades, the volume of processed and digitally stored data has reached huge dimensions. This has led to a search for innovative methods capable of coping with large data volumes. A naturally analogous context is that of intelligent information processing by higher living organisms. They are supplied by a continuous stream of voluminous sensor data (delivered by senses such as vision, hearing, or tactile sense) and use this stream for immediate or delayed acting favorable to the organism. This fact makes the field of *Artificial Intelligence* (AI) a natural source of potential ideas for Data Science. These technologies complement the findings and methods developed by classical disciplines concerned with data analysis, the most prominent of which is statistics.

The research subject of *Artificial Intelligence* (AI) is all aspects of sensing, recognition, and acting necessary for intelligent or autonomous behavior. The scope of Data Science is similar but focused on the aspects of recognition. Given the data, collected by sensing or by other data accumulation processes, the Data Science tasks consist in recognizing patterns interesting or important in some defined sense. More concretely, these tasks can adopt the form of the following variants (but not limited to them):

- recognizing one of the predefined classes of patterns (a classification task). An example is recognition of an object in a visual scene characterized by image pixel data or determining the semantic meaning of an ambiguous phrase;
- finding a quantitative relationship between some data (a continuous mapping). Such relationships are frequently found in technical and economic data, for example, the dependence of interest rate on the growth rate of domestic product or money supply;
- finding characteristics of data that are substantially more compact than the original data (data compression). A trivial example is characterizing the data about a

T. Hrycej et al., *Mathematical Foundations of Data Science*, Texts in Computer Science, https://doi.org/10.1007/978-3-031-19074-2_1

population by an arithmetic mean or standard deviation of the weight or height of
individual persons. A more complex example is describing the image data by a set
of contour edges.

Depending on the character of the task, the data processing may be static or
dynamic. The static variant is characterized by a fixed data set in which a pattern is
to be recognized. This corresponds to the mathematical concept of a mapping: Data
patterns are mapped to their pattern labels. Static recognition is a widespread setting
for image processing, text search, fraud detection, and many others.

With dynamic processing, the recognition takes place on a stream of data provided
continuously in time. The pattern searched can be found only by observing this stream
and its dynamics. A typical example is speech recognition.

Historically, the first approaches to solving these tasks date back to several cen-
turies ago and have been continually developed. The traditional disciplines have
been statistics as well as systems theory investigating dynamic system behavior.
These disciplines provide a large pool of scientifically founded findings and methods. Their natural focus on linear systems results from the fact that these systems are
substantially easier to treat analytically. Although some powerful theory extensions
to nonlinear systems are available, a widespread approach is to treat the nonlinear
systems as locally linear and use linear theory tools.

AI has passed several phases. Its origins in the 1950s focused on simple learning principles, mimicking basic aspects of the behavior of biological neuron cells.
Information to be processed has been represented by real-valued vectors. The corresponding computing procedures can be counted to the domain of numerical mathematics. The complexity of algorithms has been limited by the computing power of
information processing devices available at that time. The typical tasks solved have
been simple classification problems encompassing the separation of two classes.

Limitations of this approach with the given information processing technology
have led to an alternative view: logic-based AI. Instead of focusing on sensor information, logical statements, and correspondingly, logically sound conclusions have been
investigated. Such data has been representing some body of knowledge, motivating
to call the approach *knowledge based*. The software systems for such processing
have been labeled "expert systems" because of the necessity of encoding expert
knowledge in an appropriate logic form.

This field has reached a considerable degree of maturity in machine processing of
logic statements. However, the next obstacle had to be surmounted. The possibility of
describing a real world in logic terms showed its limits. Many relationships important
for intelligent information processing and behavior turned out to be too diffuse for
the unambiguous language of logic. Although some attempts to extend the logic
by probabilistic or pseudo-probabilistic attributes (*fuzzy logic*) delivered applicable
results, the next change of paradigm has taken place.

With the fast increase of computing power, also using interconnected computer
networks, the interest in the approach based on numerical processing of real-valued
data revived. The computing architectures are, once more, inspired by neural systems
of living organisms. In addition to the huge growth of computing resources, this phase

is characterized by more complex processing structures. Frequently, they consist of a stack of multiple subsequent processing layers. Such computing structures are associated with the recently popular notion of *Deep Learning* (DL).

The development of the corresponding methods has mostly been spontaneous and application driven. It has also taken place in several separate scientific communities, depending on their respective theoretical and application focus: computer scientists, statisticians, biologists, linguists as well as engineers of systems for image processing, speech processing, and autonomous driving. For some important applications such as *Natural Language Processing* (NLP) or *Computer Vision* (CV), many trials for solutions have been undertaken followed by equally numerous failures.

It would be exaggerated to characterize the usual approach as a "trial-and-error" approach. However, so far, no unified theory of the domain has been developed. Also, some popular algorithms and widely accepted recommendations for their use have not reached the maturity in implementation and theoretical foundations. This motivates the need for a review of mathematical principles behind the typical Data Science solutions, for the user to be able to make appropriate choices and to avoid failures caused by typical pitfalls.

Such a basic review is done in the following chapters of this work. Rather than attempting to provide a theory of *Data Science* (DS) (which would be a very ambitious project), it compiles mathematical concepts useful in looking for DS solutions. These mathematical concepts are also helpful in understanding which configurations of data and algorithms have the best chance for success. Rather than presenting a long list of alternative methods, the focus is rather on choices common to many algorithms.

What is adopted is the view of someone facing a new DS application. The questions that immediately arise are as follows:

- What type of generic task is this (forecast or classification, static or dynamic system, etc.)?
- Which are the requirements on appropriate data concerning their choice and quantity?
- Which are the conditions for generalization to unseen cases and their consequences for dimensioning the task?
- Which algorithms have the largest potential for good solutions?

The authors hope to present concise and transparent answers to these questions wherever allowed by the state of the art.

Part I
Mathematical Foundations

Application-Specific Mappings and Measuring the Fit to Data

Information processing algorithms consist of receiving input data and computing output data from them. On a certain abstraction level, they can be generally described by some kind of mapping input data to output data. Depending on software type and application, this can be more or less explicit. A dialogue software receives its input data successively and sometimes in dependence on previous input and output. By contrast, the input of a weather forecast software is a set of measurements from which the forecast (the output data) is computed by a mathematical algorithm. The latter case is closer to the common idea of a mapping in the sense of a mathematical function that delivers a function value (possibly a vector) from a vector of arguments. Depending on the application, it may be appropriate to call the input and output vectors *patterns*.

In this sense, most DS applications amount to determining some mapping of an input pattern to an output pattern. In particular, the DS approach is gaining this mapping in an inductive manner from large data sets.

Both input and output patterns are typically described by vectors. What is sought is a vector mapping

$$y = f(x) \tag{2.1}$$

assigning an output vector y to a vector x.

This mapping may be arbitrarily complex but some types are easily tractable while others are more difficult. The simplest type of mapping is linear:

$$y = Bx \tag{2.2}$$

Linear mapping is the type the most thoroughly investigated, providing voluminous theory concerning its properties. Nevertheless, its limitations induced the interest in nonlinear alternatives, the recent one being neural networks with growing popularity and application scope.

The approach typical for DS is looking for a mapping that fits to the data from some data collection. This fitting is done with the help of a set of variable parameters whose values are determined so that the fit is the best or even exact. Every mapping of type (2.1) can be written as

$$y = f(x, w) \tag{2.3}$$

© The Author(s), under exclusive license to Springer Nature Switzerland AG 2023
T. Hrycej et al., *Mathematical Foundations of Data Science*, Texts in Computer Science,
https://doi.org/10.1007/978-3-031-19074-2_2

with a parameter vector w. For linear mappings of type (2.2), the parameter vector w consists of the elements of matrix B.

There are several basic application types with their own interpretation of the mapping sought. The task of fitting a mapping of a certain type to the data requires a measure of how good this fit is. An appropriate definition of this measure is important for several reasons:

- In most cases, a perfect fit with no deviation is not possible. To select from alternative solutions, comparing the values of fit measure is necessary.
- For optimum mappings of a simple type such as linear ones, analytical solutions are known. Others can only be found by numerical search methods. To control the search, repeated evaluation of the fit measure is required.
- The most efficient search methods require smooth fit measures with existing or even continuous gradients, to determine the search direction where the chance for improvement is high.

For some mapping types, these two groups of requirements are difficult to meet in a single fit measure.

There are also requirements concerning the correspondence of the fit measure appropriate from the viewpoint of the task on one hand and of that used for (mostly numerical) optimization on the other hand:

- The very basic requirement is that both fit measures should be the same. This seemingly trivial requirement may be difficult to satisfy for some tasks such as classification.
- It is desirable that a perfect fit leads to a zero minimum of the fit measure. This is also not always satisfied, for example, with likelihood-based measures. Difficulties to satisfy these requirements frequently lead to using different measures for the search on one hand and for the evaluation of the fit on the other hand. In such cases, it is preferable if both measures have at least a common optimum.

These are the topics of the following sections.

2.1 Continuous Mappings

The most straightforward application type is using the mapping as what it mathematically is: a mapping of real-valued input vectors to equally real-valued output vectors. This type encompasses many physical, technical, and econometric applications. Examples of this may be:

Fig. 2.1 Error functions

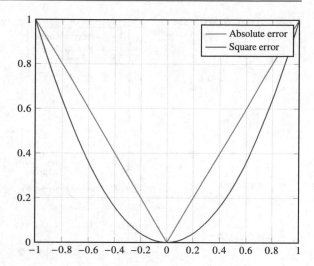

- Failure rates (y) determined from operation time and conditions of a component (x).
- Credit scoring, mapping the descriptive features (x) of the credit recipient to a number denoting the creditworthiness (y).
- Macroeconomic magnitudes such as inflation rate (y) estimated from others such as unemployment rate and economic growth (x).

If a parameterized continuous mapping is to be fitted to data, the goal of fitting is to minimize the deviation between the true values y and the estimated values $f(x, w)$. So, the basic version of fitting error is

$$e = y - f(x, w) \tag{2.4}$$

It is desirable for this error to be a small positive or negative number. In other words, it is its absolute value $|e|$ that is to be minimized for all training examples. This error function is depicted in Fig. 2.1 and labeled *absolute error*.

An obvious property of this error function is its lack of smoothness. Figure 2.2 shows its derivatives: It does not exist at $e = 0$ and makes a discontinuous step at that position $e = 0$.

This is no problem from the application's view. However, a discontinuous first derivative is strongly adverse for the most well-converging numerical algorithms that have a potential to be used as fitting or training algorithms. It is also disadvantageous for analytical treatment. The error minimum can be sometimes determined analytically, seeking for solutions with zero derivative. But equations containing discontinuous functions are difficult to solve. Anyway, the absolute value as an error measure is used in applications with special requirements such as enhanced robustness against data outliers.

Numerical tractability is the reason why a preferred form of error function is a square error e^2, also shown in Fig. 2.1. Its derivative (Fig. 2.2) is not only continuous but even linear, which makes its analytical treatment particularly easy.

Fig. 2.2 Derivatives of error functions

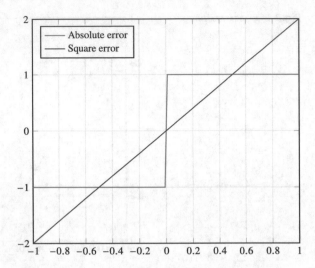

For a vector mapping $f(x, w)$, the error (2.4) is a column vector. The vector product $e'e$ is the sum of the squares of the errors of individual output vector elements. Summing these errors over K training examples result in the error measure

$$E = \sum_{k=1}^{K} e'_k e_k = \sum_{k=1}^{K} \sum_{m=1}^{M} e_{mk}^2 \tag{2.5}$$

Different scaling of individual elements of vector patterns can make scaling weights $S = [s_1 \dots s_M]$ appropriate. Also, some training examples may be more important than others, which can be expressed by additional weights r_k. The error measure (2.5) has then the generalized form

$$E = \sum_{k=1}^{K} e'_k S e_k r_k = \sum_{k=1}^{K} \sum_{m=1}^{M} e_{mk}^2 s_m r_k \tag{2.6}$$

2.1.1 Nonlinear Continuous Mappings

For linear mappings (2.2), explicit solutions for reaching zero in the error measure (2.5) and (2.6) are known. Their properties have been thoroughly investigated and some important aspects are discussed in Chap. 4. Unfortunately, most practical applications deviate to a greater or lesser extent from the linearity assumption. Good analytical tractability may be a good motivation to accept a linear approximation if the expected deviations from the linearity assumption are not excessive. However, a lot of applications will not allow such approximation. Then, some nonlinear approach is to be used.

Modeling nonlinearities in the mappings can be done in two ways that strongly differ their application.

The first approach preserves *linearity in parameters*. The mapping (2.3) is expressed as

$$y = Bh(x) \tag{2.7}$$

with a nonparametric function $h(x)$ which plays the role of the input vector x itself. In other words, $h(x)$ can be substituted for x in all algebraic relationships valid for linear systems. This includes also explicit solutions for *Mean Square Errors* (MSEs) (2.5) and (2.6).

The function $h(x)$ can be an arbitrary function but a typical choice is a polynomial in vector x. This is motivated by the well-known Taylor expansion of an arbitrary multivariate function [7]. This expansion enables an approximation of a multivariate function by a polynomial of a given order on an argument interval, with known error bounds.

For a vector x with two elements x_1 and x_2, a quadratic polynomial is

$$h\left(\begin{bmatrix} x_1 & x_2 \end{bmatrix}'\right) = \begin{bmatrix} 1 & x_1 & x_2 & x_1^2 & x_2^2 & x_1 x_2 \end{bmatrix}' \tag{2.8}$$

For a vector x with three elements x_1, x_2, and x_3, it is already as complex as follows:

$$h\left(\begin{bmatrix} x_1 & x_2 & x_3 \end{bmatrix}'\right) = \begin{bmatrix} 1 & x_1 & x_2 & x_3 & x_1^2 & x_2^2 & x_3^2 & x_1 x_2 & x_1 x_3 & x_2 x_3 \end{bmatrix}' \tag{2.9}$$

For a vector x of length N, the length of vector $h(x)$ is

$$1 + N + N + \frac{(N-1)N}{2} = 1 + \frac{N^2 + 3N}{2} \tag{2.10}$$

For a polynomial of order p, the size of vector $h(x)$ grows with the pth power of N. This is the major shortcoming of the polynomial approach for typical applications of DS where input variable numbers of many thousands are common. Already with quadratic polynomials, the input width would increase to millions and more.

Another disadvantage is the growth of higher polynomial powers outside of the interval covered by the training set—a minor extrapolation may lead to excessively high output values.

So, modeling the multivariate nonlinearities represented by polynomials is practical only for low-dimensional problems or problems in which it is justified to refrain from taking full polynomials (e.g., only powers of individual scalar variables). With such problems, it is possible to benefit from the existence of analytical optima and statistically well-founded statements about the properties of the results.

These properties of parameterized mappings linear in parameters have led to the high interest in more general approximation functions. They form the second approach: *mappings nonlinear in parameters*. A prominent example are *neural networks*, discussed in detail in Chap. 3. In spite of intensive research, practical statements about their representational capacity are scarce and overly general, although there are some interesting concepts such as *Vapnik–Chervonenkis* dimension [21].

Neural networks with bounded activation functions such as *sigmoid* do not exhibit the danger of unbounded extrapolation. They frequently lead to good results if the number of parameters scales linearly with the input dimension, although the optimality or appropriateness of their size is difficult to show. Determining their optimum size is frequently a result of lengthy experiments.

Fitting neural networks to data is done numerically because of missing analytical solutions. This makes the use of well-behaving error functions such as MSE particularly important.

2.1.2 Mappings of Probability Distributions

Minimizing the MSE (2.5) or (2.6) leads to a mapping making a good (or even perfect, in the case of a zero error) forecast of the output vector y. This corresponds to the statistical concept of *point estimation* of the expected value of y.

In the presence of an effect unexplained by input variable or of some type of noise, the true values of the output will usually not be exactly equal to their expected values. Rather, they will fluctuate around these expected values according to some probability distribution. If the scope of these fluctuations is different for different input patterns x, the knowledge of the probability distribution may be of crucial interest for the application. In this case, it would be necessary to determine a conditional probability distribution of the output pattern y conditioned on the input pattern x

$$g\,(y \mid x) \tag{2.11}$$

If the expected probability distribution type is parameterized by parameter vector p, then (2.11) extends to

$$g\,(y \mid x, p) \tag{2.12}$$

From the statistical viewpoint, the input/output mapping (2.3) maps the input pattern x directly to the point estimator of the output pattern y. However, we are free to adopt a different definition: input pattern x can be mapped to the conditional parameter vector p of the distribution of output pattern y. This parameter vector has nothing in common with the fitted parameters of the mapping—it consists of parameters that determine the shape of a particular probability distribution of the output patterns y, given an input pattern x. After the fitting process, the conditional probability distribution (2.12) becomes

$$g\,(y, f\,(x, w)) \tag{2.13}$$

It is an unconditional distribution of output pattern y with distribution parameters determined by the function $f\,(x, w)$. The vector w represents the parameters of the mapping "input pattern $x \Rightarrow$ conditional probability distribution parameters p" and should not be confused with the distribution parameters p themselves. For example, in the case of mapping $f\,()$ being represented by a neural network, w would correspond to the network weights. Distribution parameters p would then correspond to the activation of the output layer of the network for a particular input pattern x.

This can be illustrated on the example of a multivariate normal distribution with a mean vector m and covariance matrix C. The distribution (2.12) becomes

$$g\,(y \mid x, p) = N\,(m\,(x)\,, C\,(x)) = \frac{1}{\sqrt{(2\pi)^N |C\,(x)|}}\,\mathrm{e}^{\frac{-1}{2}(y - m(x))' C(x)^{-1}(y - m(x))} \tag{2.14}$$

The vector y can, for example, represent the forecast of temperature and humidity for the next day, depending on today's meteorological measurements x. Since the point forecast would scarcely hit the tomorrow's state and thus be of limited use, it will be substituted by the forecast that the temperature/humidity vector is expected to have the mean $m(x)$ and the covariance matrix $C(x)$, both depending on today's measurement vector x. Both the mean vector and the elements of the covariance matrix together constitute the distribution parameter vector p in (2.12). This parameter vector depends on the vector of meteorological measurements x as in (2.13).

What remains is to choose an appropriate method to find the optimal mappings $m(x)$ and $C(x)$ which depend on the input pattern x. In other words, we need some optimality measure for the fit, which is not as simple as in the case of point estimation with its square error. The principle widely used in statistics is that of *maximum likelihood*. It consists of selecting distribution parameters (here: m and C) such that the probability density value for the given data is maximum.

For a training set pattern pair (x_k, y_k), the probability density value is

$$\frac{1}{\sqrt{(2\pi)^N |C(x_k)|}} e^{-\frac{1}{2}(y_k - m(x_k))' C(x_k)^{-1}(y_k - m(x_k))} \tag{2.15}$$

For independent samples (x_k, y_k), the likelihood of the entire training set is the product

$$\prod_{k=1}^{K} \frac{1}{\sqrt{(2\pi)^N |C(x_k)|}} e^{-\frac{1}{2}(y_k - m(x_k))' C(x_k)^{-1}(y_k - m(x_k))} \tag{2.16}$$

The maximum likelihood solution consists in determining the mappings $m(x)$ and $C(x)$ such that the whole product term (2.16) is maximum.

The exponential term in (2.16) suggests that taking the logarithm of the expression may be advantageous, converting the product to a sum over training patterns. A negative sign would additionally lead to a minimizing operation, consistently with the convention for other error measures which are usually minimized. The resulting term

$$\sum_{k=1}^{K} \left(\frac{N}{2} \ln 2\pi + \frac{1}{2} \ln |C(x_k)| + \frac{1}{2} (y_k - m(x_k))' C(x_k)^{-1} (y_k - m(x_k)) \right) \tag{2.17}$$

can be simplified by rescaling and omitting constants to

$$\sum_{k=1}^{K} \left(\ln |C(x_k)| + (y_k - m(x_k))' C(x_k)^{-1} (y_k - m(x_k)) \right) \tag{2.18}$$

In the special case of known covariance matrices $C(x_k)$ independent from the input pattern x_k, the left term is a sum of constants and (2.18) reduces to a generalized MSE, an example of which is the measure (2.6). The means $m(x_k)$ are then equal to the point estimates received by minimizing the MSE.

More interesting is the case of unknown conditional covariance matrices $C(x_k)$. Its form advantageous for computations is based on the following algebraic laws:

- Every symmetric matrix such as $C(x_k)^{-1}$ can be expressed as a product of a lower triangular matrix L and its transpose L', that is, $C(x_k)^{-1} = L(x_k) L(x_k)'$.
- The determinant of a lower diagonal matrix L is the product of its diagonal elements.
- The determinant of LL' is the square of the determinant of L.
- The inverse L^{-1} of a lower diagonal matrix L is a lower diagonal matrix and its determinant is the reciprocal value of the determinant of L.

The expression (2.18) to be minimized becomes

$$\sum_{k=1}^{K} \left(-2 \sum_{m=1}^{M} (\ln(l_{mm}(x_k))) + (y_k - m(x_k))' L(x_k) L(x_k)' (y_k - m(x_k)) \right)$$

(2.19)

The mapping $f(x, w)$ of (2.13) delivers for every input pattern x_k the mean vector $m(x_k)$ and the lower triangular matrix $L(x_k)$ of structure

$$\begin{bmatrix} l_{11} & \cdots & 0 \\ \vdots & \ddots & \vdots \\ l_{M1} & \cdots & l_{MM} \end{bmatrix}$$

(2.20)

So, for input pattern x_k, the output vector y_k is forecast to have the distribution

$$\frac{1}{\sqrt{(2\pi)^N \prod_{m=1}^{M} l_{mm}^{-2}(x_k)}} e^{\frac{-1}{2}(y_k - m(x_k))' L(x_k) L(x_k)' (y_k - m(x_k))}$$

(2.21)

If the mapping $f(x, w)$ is represented, for example, by a neural network, the output layer of the network is trained to minimize the log-likelihood (2.19), with the tuple $(m(x_k), L(x_k))$ extracted from the corresponding elements of the output layer activation vector.

For higher dimensions M of output pattern vector y, the triangular matrix L has a number of entries growing with the square of M. Only if mutual independence of individual output variables can be assumed, L becomes diagonal and has a number of nonzero elements equal to M.

There are few concepts alternative to multivariate normal distribution if mutual dependencies are essential (in our case, mutual dependencies within the output pattern vector y). A general approach has been presented by Stützle and Hrycej [19]. However, if the independence assumption is justified, arbitrary univariate distributions can be used with specific parameters for each output variable y_m. For example, for modeling the time to failure of an engineering component, the *Weibull* distribution [22] is frequently used, with the density function

$$g(y) = \frac{\beta}{\eta} \left(\frac{y}{\eta} \right)^{\beta-1} e^{-\left(\frac{y}{\eta} \right)^{\beta}}$$

(2.22)

We are then seeking the parameter pair $(\beta(x), \eta(x))$ depending on the input pattern x such that the log-likelihood over the training set

$$\sum_{k=1}^{K} \ln \frac{\beta(x_k)}{\eta(x_k)} + \beta(x_k) - \ln \frac{y_k}{\eta(x_k)} - \left(\frac{y_k}{\eta(x_k)}\right)^{\beta(x_k)}$$

$$= \sum_{k=1}^{K} \ln \beta(x_k) - \beta(x_k) \ln \eta(x_k) + (\beta(x_k) - 1) \ln y_k - \left(\frac{y_k}{\eta(x_k)}\right)^{\beta(x_k)} \tag{2.23}$$

is minimum. The parameter pair can, for example, be the output layer (of size 2) activation vector

$$\begin{bmatrix} \beta & \eta \end{bmatrix}' = f(x, w) \tag{2.24}$$

2.2 Classification

A classification problem is characterized by assigning every pattern a class out of a predefined class set. Such problems are frequently encountered whenever the result of a mapping is to be assigned some verbal category. Typical examples are

- images in which the object type is sought (e.g., a face, a door, etc.);
- radar signature assigned to flying objects;
- object categories on the road or in its environment during autonomous driving.

Sometimes, the classes are only discrete substitutes for a continuous scale. Discrete credit scores such as "fully creditworthy" or "conditionally creditworthy" are only distinct values of a continuous variable "creditworthiness score". Also, many social science surveys classify the answers to "I fully agree", "I partially agree", "I am indifferent", "I partially disagree", and "I fully disagree", which can be mapped to a continuous scale, for example $[-1, 1]$. Generally, this is the case whenever the classes can be ordered in an unambiguous way.

Apart from this case with inherent continuity, the classes may be an order-free set of exclusive alternatives. (Nonexclusive classifications can be viewed as separate tasks—each nonexclusive class corresponding to a dichotomy task "member" vs. "nonmember".) For such class sets, a basic measure of the fit to a given training or test set is the misclassification error. The misclassification error for a given pattern may be defined as a variable equal to zero if the classification by the model corresponds to the correct class and equal to one if it does not. More generally, assigning the object with the correct class i erroneously to the class j is evaluated by a nonnegative real number called *loss* L_{ij}. The loss of a correct class assignment is $L_{ii} = 0$.

The so-defined misclassification loss is a transparent measure, frequently directly reflecting application domain priorities. By contrast, it is less easy to make it operational for fitting or learning algorithms. This is due to its discontinuous character—a class assignment can only be correct or wrong. So far, solutions have been found only for special cases.

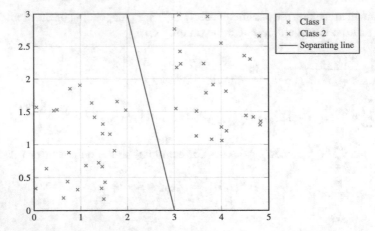

Fig. 2.3 Two classes with linear separation

This discontinuity represents a difficulty when searching for optimal classification mappings. For continuous mappings, there is a comfortable situation that a fit measure such as MSE is also one that can be directly used in numerical optimization. Solving the optimization problem is identical with solving the fitting problem. For classification tasks, this comfort cannot be enjoyed. What we want to reach is not always identical with what we can efficiently optimize. To bridge this gap, various approaches have been proposed. Some of them are sketched in the following sections.

2.2.1   Special Case: Two Linearly Separable Classes

Let us consider a simple problem with two classes and two-dimensional patterns $[x_1, x_2]$ as shown in Fig. 2.3. The points corresponding to Class 1 and Class 2 can be completely separated by a straight line, without any misclassification. This is why such classes are called *linearly separable*. The attainable misclassification error is zero.

The existence of a separating line guarantees the possibility to define regions in the pattern vector space corresponding to individual classes. What is further needed is a function whose value would indicate the membership of a pattern in a particular class. Such function for the classes of Fig. 2.3 is that of Fig. 2.4. Its value is unity for patterns from Class 1 and zero for those from Class 2.

Unfortunately, this function has properties disadvantageous for treatment by numerical algorithms. It is discontinuous along the separating line and has zero gradient elsewhere. This is why it is usual to use an indicator function of type shown in Fig. 2.5. It is a linear function of the pattern variables. The patterns are assigned to Class 1 if this function is positive and to Class 2 otherwise.

Many or even the most class pairs cannot be separated by a linear hyperplane. It is not easy to determine whether they can be separated by an arbitrary function if the

Fig. 2.4 Separating function

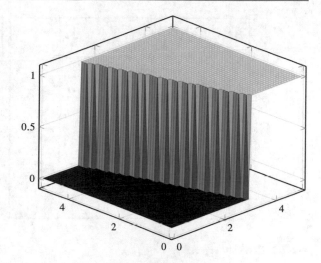

Fig. 2.5 Continuous
separating function

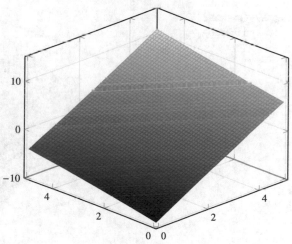

family of these functions is not fixed. However, some classes can be separated by simple surfaces such as quadratic ones. An example of this is given in Fig. 2.6. The separating curve corresponds to the points where the separating function of Fig. 2.7 intersects the plane with $y = 0$.

The discrete separating function such as that of Fig. 2.4 can be viewed as a nonlinear step function of the linear function of Fig. 2.5, that is,

$$s\left(b'x\right) = \begin{cases} 1 & \text{for } b'x \geq 0 \\ 0 & \text{for } b'x < 0 \end{cases} \tag{2.25}$$

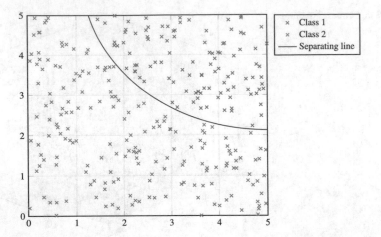

Fig. 2.6 Quadratic separation

Fig. 2.7 Quadratic
separation function

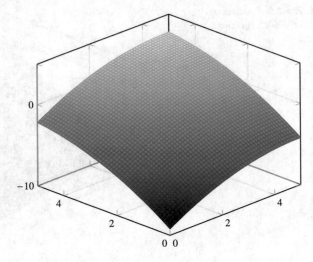

To avoid explicitly mentioning the absolute term, it will be assumed that the last
element of input pattern vector x is equal to unity, so that

$$b'x = \begin{bmatrix} b_1 & \cdots & b_{N-1} & b_N \end{bmatrix} \begin{bmatrix} x_1 \\ \vdots \\ x_{N-1} \\ 1 \end{bmatrix} = \begin{bmatrix} b_1 & \cdots & b_{N-1} \end{bmatrix} \begin{bmatrix} x_1 \\ \vdots \\ x_{N-1} \end{bmatrix} + b_N$$

The misclassification sum for a training set with input/output pairs (x_k, y_k) is
equal to

$$E = \sum_{k=1}^{K} \left(s \left(b' x_k \right) - y_k \right)^2 \tag{2.26}$$

Here, y_k is the class indicator of the kth training pattern with values 0 or 1. For most numerical minimization methods for error functions E, the gradient of E with regard to parameters b is required to determine the direction of descent towards low values of E. The gradient is

$$\frac{\partial E}{\partial b} = 2 \sum_{k=1}^{K} \left(s\left(b'x_k\right) - y_k \right) \frac{\mathrm{d}s}{\mathrm{d}z} x_k \tag{2.27}$$

with z being the argument of function $s(z)$.

However, the derivative of the nonlinear step function (2.25) is zero everywhere except for the discontinuity at $z = 0$ where it does not exist. To receive a useful descent direction, the famous *perceptron rule* [16] has used a gradient modification. This pioneering algorithm iteratively updates the weight vector b in the direction of the (negatively taken) modified gradient

$$\frac{\partial E}{\partial b} = \sum_{k=1}^{K} \left(s\left(b'x_k\right) - y_k \right) x_k \tag{2.28}$$

This modified gradient can be viewed as (2.27) with $\frac{\mathrm{d}s}{\mathrm{d}z}$ substituted by unity (the derivative of linear function $s(z) = z$). Taking a continuous gradient approximation is an idea used by optimization algorithms for non-smooth functions, called *subgradient algorithms* [17].

The algorithm using the perceptron rule converges to zero misclassification rate if the classes, as defined by the training set, are separable. Otherwise, convergence is not guaranteed.

An error measure focusing on critical patterns in the proximity of separating line is used by the approach called the *support vector machine* (SVM) [2]. This approach is looking for a separating line with the largest orthogonal distance to the nearest patterns of both classes. In Fig. 2.8, the separating line is surrounded by the corridor defined by two boundaries against both classes, touching the respective nearest points. The goal is to find a separating line for which the width of this corridor is the largest. In contrast to the class indicator of Fig. 2.4 (with unity for Class 1 and zero for Class 2), the support vector machine rule is easier to represent with a symmetric class indicator y equal to 1 for one class and to -1 for another one. With this class indicator and input pattern vector containing the element 1 to provide for the absolute bias term, the classification task is formulated as a constrained optimization task with constraints

$$y_k b' x_k \geq 1 \tag{2.29}$$

If these constraints are satisfied, the product $b'x_k$ is always larger than 1 for Class 1 and smaller than -1 for Class 2.

The separating function $b'x$ of (2.29) is a hyperplane crossing the x_1/x_2-coordinates plane at the separating line (red line in Fig. 2.8). At the boundary lines, $b'x$ is equal to constants larger than 1 (boundary of Class 1) and smaller than -1 (boundary of Class 2). However, there are infinitely many such separating functions. In the

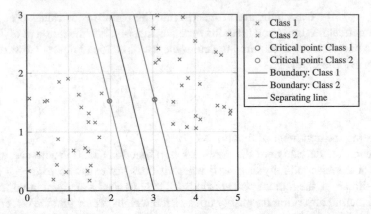

Fig. 2.8 Separating principle of a SVM

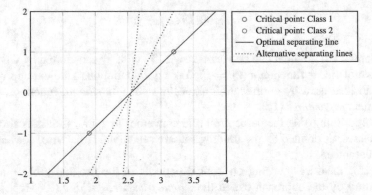

Fig. 2.9 Alternative separating functions—cross-sectional view

cross section perpendicular to the separating line (i.e., viewing the x_1/x_2-coordinates plane "from aside"), they may appear as in Fig. 2.9.

There are infinitely many such hyperplanes (appearing as dotted lines in the cross section of Fig. 2.9), some of which becoming very "steep". The most desirable variant would be that exactly touching the critical points of both classes at a unity "height" (solid line). This is why the optimal solution of the SVM is such that it has the minimum norm of vector b:

$$\min \|b\| \tag{2.30}$$

The vector norm is a quadratic function of vector elements. So, the constraints (2.29) together with the objective function (2.30) constitute a quadratic minimization problem with constraints, solvable with modern numerical methods. Usually, the dual form having the same optimum as the problem in (2.29) and (2.30) is solved.

Both the perceptron rule and the SVM are originally designed for linearly separable classes. In this case, the optimum corresponds to the perfect separation and no misclassification occurs. With linearly separable classes, the measure of success is simple: "separated" (successful fit) and "non-separated" (failing to fit). The absence of intermediary results makes the problem of discontinuous misclassification error or loss irrelevant—every separation is a full success.

2.2.2 Minimum Misclassification Rate for Two Classes

Unfortunately, separability or even linear separability is rather scarce in real-world classification problems. Then, the minimization of the inevitable misclassification loss is the genuine objective. The perceptron rule and the SVM have extensions for non-separable classes but they do not perform genuine misclassification minimization although the results may be acceptable.

The group of methods explicitly committed to this goal are found in the statistical discriminant analysis. The principle behind the typically applied approach is to determine the probability of a pattern vector to be a member of a certain class. If these probabilities are known (or estimated in a justified way) for all classes in questions, it is possible to choose the class with the highest probability. If the probability that a pattern is a member of the ith class is P_i, the probability of being assigned a false class $j \neq i$ is $1 - P_i$. If every pattern is assigned to the class with the highest probability, the probability of misclassification (which is proportional to the misclassification error) is at its minimum.

With the knowledge of a probability distribution of the patterns of each class, this assessment can be made. In other words, the classification problem is "reduced" to the task of assessment of these probability distributions for all classes. The quotation marks around "reduced" suggest that this task is not easy. On the contrary, it is a formidable challenge since most real-world pattern classes follow no analytical distribution used in the probability theory.

Let us consider the case of two classes, the patterns of each of which are normally distributed (Gaussian distribution), with mean vector m_i, covariance matrix C_i, $i = 1, 2$, and pattern vector length N:

$$N(m_i, C_i) = \frac{1}{\sqrt{(2\pi)^N |C_i|}} e^{\frac{-1}{2}(x-m_i)'C_i^{-1}(x-m_i)} \tag{2.31}$$

The density (2.31) can be viewed as a conditional density $f(x \mid i)$ given the class i. The classes may have different prior probabilities p_i (i.e., they do not occur equally frequently in reality). Bayesian posterior probability of pattern x being the ith class is then

$$P_i = \frac{f(x \mid i) p_i}{f(x \mid 1) p_1 + f(x \mid 2) p_2} \tag{2.32}$$

Which class has a higher probability can be tested by comparing the ratio

$$\frac{P_1}{P_2} = \frac{f(x \mid 1) p_1}{f(x \mid 2) p_2} \tag{2.33}$$

with unity, or, alternatively, comparing its logarithm

$$\ln(P_1) - \ln(P_2) = \ln(f(x \mid 1)) - \ln(f(x \mid 2)) + \ln(p_1) + \ln(p_2) \tag{2.34}$$

with zero.

Substituting (2.31)–(2.34) results in

$$
\ln\left(\frac{1}{\sqrt{(2\pi)^N|C_1|}}\right) - \frac{1}{2}(x-m_1)'C_1^{-1}(x-m_1)
$$

$$
- \ln\left(\frac{1}{\sqrt{(2\pi)^N|C_2|}}\right) + \frac{1}{2}(x-m_2)'C_2^{-1}(x-m_2)
$$

$$
+ \ln(p_1) - \ln(p_2) \tag{2.35}
$$

$$
= \frac{1}{2}\ln(|C_2|) - \frac{1}{2}\ln(|C_1|) + \frac{1}{2}x'\left(C_2^{-1} - C_1^{-1}\right)x
$$

$$
+ \left(m_1'C_1^{-1} - m_2'C_2^{-1}\right)x - \frac{1}{2}\left(m_1'C_1^{-1}m_1 - m_2'C_2^{-1}m_2\right)
$$

$$
+ \ln(p_1) - \ln(p_2)
$$

which can be made more transparent as

$$
x'Ax + b'x + d \tag{2.36}
$$

with

$$
A = \frac{1}{2}C_2^{-1} - C_1^{-1}
$$

$$
b' = m_1'C_1^{-1} - m_2'C_2^{-1}
$$

$$
d = \frac{1}{2}\ln(|C_2|) - \frac{1}{2}\ln(|C_1|) - \frac{1}{2}\left(m_1'C_1^{-1}m_1 - m_2'C_2^{-1}m_2\right) + \ln(p_1) - \ln(p_2)
$$

A Bayesian optimum decision consists in assigning the pattern to `Class 1` if the expression (2.36) is positive and to `Class 2` if it is negative.

Without prior probabilities p_i, the ratio (2.33) is the so-called *likelihood ratio* which is a popular and well elaborated statistical decision criterion. The decision function (2.36) is then the same, omitting the logarithms of p_i.

The criterion (2.36) is a quadratic function of pattern vector x. The separating function is of type depicted in Fig. 2.7. This concept can be theoretically applied to some other distributions beyond the Gaussian [12].

The discriminant function (2.36) is dedicated to classes with normally distributed classes. If the mean vectors and the covariance matrices are not known, they can easily be estimated from the training set, as sample mean vectors and sample covariance matrices, with well-investigated statistical properties. However, the key problem is the assumption of normal distribution itself. It is easy to imagine that this assumption is rather scarcely strictly satisfied. Sometimes, it is even clearly wrong.

Practical experience has shown that the discriminant function (2.36) is very sensitive to deviations from distribution normality. Paradoxically, better results are usually reached with a further assumption that is even less frequently satisfied: that of a common covariance matrix C identical for both classes. This is roughly equivalent to the same "extension" of both classes in the input vector space.

For Gaussian classes with column vector means m_1 and m_2, and common co-variance matrix C, matrix A and some parts of the constant d become zero. The discriminant function becomes linear:

$$b'x + d > 0$$

with

$$b = (m_1 - m_2)'C^{-1}$$

$$d = -\frac{1}{2}b'(m_1 + m_2) + \ln\frac{p_1}{p_2}$$

$$= -\frac{1}{2}(m_1 + m_2)'C^{-1}(m_1 + m_2) + \ln\frac{p_1}{p_2}$$

(2.37)

This linear function is widely used in the *linear discriminant analysis*.

Interestingly, the separating function (2.37) can, under some assumptions, be received also with a least squares approach. For simplicity, it will be assumed that the mean over both classes $m_1 p_2 + m_2 p_2$ is zero. Class 1 and Class 2 are coded by 1 and -1, and the pattern vector x contains 1 at the last position.

The zero gradient is reached at

$$b'XX' = yX'$$

(2.38)

By dividing both sides by the number of samples, matrices X and XX' contain sample moments (means and covariances). Expected values are

$$E\left[\frac{1}{K}b'XX'\right] = E\left[\frac{1}{K}yX'\right]$$

(2.39)

The expression XX' corresponds to the sample second moment matrix. With the zero mean, as assumed above, it is equal to the sample covariance matrix. Every covariance matrix over a population divided into classes can be decomposed to the intraclass covariance C (in this case, identical for both classes) and the interclass covariance

$$M = \begin{bmatrix} m_1 & m_2 \end{bmatrix}$$

$$P = \begin{bmatrix} p_1 & 0 \\ 0 & p_2 \end{bmatrix}$$

(2.40)

$$C_{cl} = MPM'$$

This can be then rewritten as

$$b'\begin{bmatrix} C + MPM' & 0 \\ 0 & 1 \end{bmatrix} = \begin{bmatrix} p_1 m_1' - p_2 m_2' & p_1 - p_2 \end{bmatrix}$$

(2.41)

resulting in

$$b' = \begin{bmatrix} p_1 m_1' - p_2 m_2' & p_1 - p_2 \end{bmatrix}\begin{bmatrix} C + C_{cl} & 0 \\ 0 & 1 \end{bmatrix}^{-1}$$

$$= \begin{bmatrix} p_1 m_1' - p_2 m_2' & p_1 - p_2 \end{bmatrix}\begin{bmatrix} [C + C_{cl}]^{-1} & 0 \\ 0 & 1 \end{bmatrix}$$

(2.42)

$$= \begin{bmatrix} (p_1 m_1' - p_2 m_2')[C + C_{cl}]^{-1} & p_1 - p_2 \end{bmatrix}$$

It is interesting to compare the linear discriminant (2.37) with least square solution (2.37) and (2.42). With an additional assumption of both classes having identical prior probabilities $p_1 = p_2$ (and identical counts in the training set), the absolute term of both (2.37) and (2.42) becomes zero. The matrix C_{cl} contains covariances of only two classes and is thus of maximum rank two. The additional condition of overall mean equal to zero reduces the rank to one. This results in least squares-based separating vector b to be only rescaled in comparison with that of separating function (2.37). This statement can be inferred in the following way.

In the case of identical prior probabilities of both classes, the condition of zero mean of distribution of all patterns is $m_1 + m_2 = 0$, or $m_2 = -m_1$. It can be rewritten as $m_1 = m$ and $m_2 = -m$ with the help of a single column vector of class means m. The difference of both means is $m_1 - m_2 = 2m$. The matrix C_{cl} is

$$C_{cl} = \frac{1}{2} [m_1 \ m_2] [m_1 \ m_2]' = \frac{1}{2} (m_1 m_1' + m_2 m_2') = mm' \qquad (2.43)$$

with rank equal to one—it is an outer product of only one vector m with itself.

The equation for separating function b of the linear discriminant is

$$bC = 2m' \qquad (2.44)$$

while for separating function b_{LS} of least squares, it is

$$b_{LS} (C + C_{cl}) = 2m' \qquad (2.45)$$

Let us assume the proportionality of both solutions by factor d:

$$b_{LS} = db \qquad (2.46)$$

Then

$$db (C + C_{cl}) = 2dm' + 2dm' C^{-1} C_{cl} = 2m' \qquad (2.47)$$

or

$$m' C^{-1} C_{cl} = m' C^{-1} mm' = \frac{1-d}{d} m' = em' \qquad (2.48)$$

with

$$e = \frac{1-d}{d} \qquad (2.49)$$

and

$$d = \frac{1}{1+e} \qquad (2.50)$$

The scalar proportionality factor e in (2.48) can always be found since $C_{cl} = mm'$ is a projection operator to a one-dimensional space. It projects every vector, i.e., also the vector $m'C^{-1}$, to the space spanned by vector m. In other words, these two vectors ale always proportional. Consequently, a scalar proportionality factor d for separating functions can always be determined via (2.50). This means that proportional separating functions are equivalent since they separate identical regions.

The result of this admittedly tedious argument is that the least square solution fitting the training set to the class indicators 1 and -1 is equivalent with the optimum linear discriminant, under the assumption of

Fig. 2.10 Lateral projection of a linear separating function (class indicators 1 and −1)

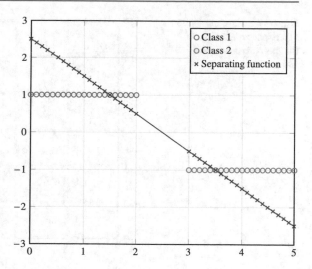

- normally distributed classes;
- identical covariance matrix of both classes;
- and classes with identical prior probabilities.

This makes the least squares solution interesting since it can be applied without assumptions about the distribution—of course with the caveat that is not Bayesian optimal for other distributions. This seems to be the foundation of the popularity of this approach beyond the statistical community, for example, in neural network-based classification.

Its weakness is that the MSE reached cannot be interpreted in terms of misclassification error—we only know that in the MSE minimum, we are close to the optimum separating function. The reason for this lack of interpretability is that the function values of the separating function are growing with the distance from the hyperplane separating both classes while the class indicators (1 and −1) are not—they remain constant at any distance. Consequently, the MSE attained by optimization may be large even if the classes are perfectly separated. This can be seen if imagining a "lateral view" of the vector space given in Fig. 2.10. It is a cross section in the direction of the class separating line. The class indicators are constant: 1 (Class 1 to the left) and −1 (Class 2 to the right).

More formally, the separating function (for the case of separable classes) assigns the patterns, according to the test $b'x + d > 0$ for Class 1 membership, to the respective correct class. However, the value of $b'x + d$ is not equal to the class indicator y (1 or −1). Consequently, the MSE $(b'x + d - y)^2$ is far away from zero in the optimum. Although alternative separating functions with identical separating lines can have different slopes, no one of them can reach zero MSE. So, the MSE does not reflect the misclassification rate.

This shortcoming can be alleviated by using a particular nonlinear function of the term $b'x + d$. Since this function is usually used in the form producing class

Fig. 2.11 Lateral projection of a linear separating function (class indicators 1 and 0)

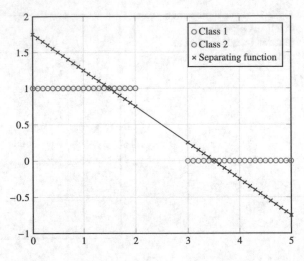

Fig. 2.12 Logistic (sigmoid) function

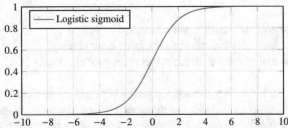

indicators 1 for `Class 1` and zero for `Class 2`, it will reflect the rescaled linear situation of Fig. 2.11.

The nonlinear function is called *logistic* or *logit function* in statistics and econometrics. With neural networks, it is usually referred to as *sigmoid function*, related via rescaling to *tangent hyperbolicus* (tanh). It is a function of scalar argument z:

$$y = s(z) = \frac{1}{1 + e^{-z}} \tag{2.51}$$

This function is mapping the argument $z \in (-\infty, \infty)$ to the interval $[0, 1]$, as shown in Fig. 2.12.

Applying (2.51) to the linear separating function $b'x + d$, that is, using the nonlinear separating function

$$y = s(b'x + d) = \frac{1}{1 + e^{-(b'x+d)}} \tag{2.52}$$

will change the picture of Fig. 2.11 to that of Fig. 2.13. The forecast class indicators (red crosses) are now close to the original ones (blue and green circles).

The MSE is

$$\left(s(b'x + d) - y\right)^2 \tag{2.53}$$

For separable classes, MSE can be made arbitrarily close to zero, as depicted in Fig. 2.14. The proximity of the forecast and true class indicators can be increased

Fig. 2.13 Lateral projection of a logistic separating function (class indicators 1 and 0)

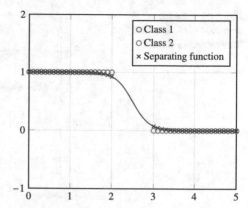

Fig. 2.14 A sequence of logistic separating functions

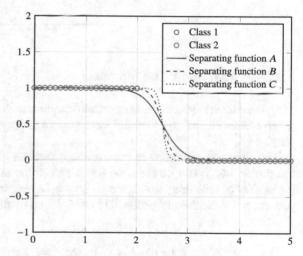

by increasing the norm of weight vector b. This unbounded norm is a shortcoming of this approach if used with perfectly separable classes.

For non-separable classes, this danger disappears. There is an optimum in which the value of the logistic class indicator has the character of probability. For patterns in the region where the classes overlap, the larger its value, the more probable is the membership in Class 1. This is illustrated as the lateral projection in Fig. 2.15.

Unfortunately, in contrast to a linear separating function and Gaussian classes, minimizing the MSE with the a logistic separating function has no guaranteed optimality properties with regard to neither misclassification loss nor class membership probability.

How to follow the probabilistic cue more consequently is discussed in the following Sect. 2.2.3.

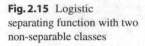

Fig. 2.15 Logistic
separating function with two
non-separable classes

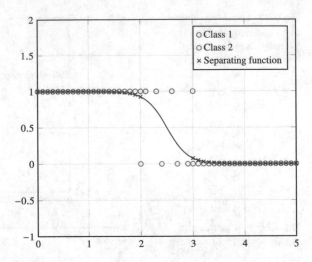

2.2.3 Probabilistic Classification

First, it has to be mentioned that misclassification rate itself is a probabilistic concept.
It can be viewed as the probability that a pattern is erroneously assigned to a wrong
class.

The approach discussed in this section adopts another view. With two classes,
the probability p can be assessed that a pattern belongs to Class 1 while the
probability of belonging to Class 2 is complementary, that is $1 - p$. For a given
pattern, p is a conditional probability conditioned by this pattern:

$$p(x) = P(y = 1 \mid x) \tag{2.54}$$

From the probabilistic point of view, the class membership of pattern x is a random process, governed by Bernoulli distribution—a distribution with exactly the
properties formulated above: probability p of membership in Class 1 and $1 - p$
for the opposite. The probability is a function of input pattern vector x.

The classification problem consists in finding a function $f(x, w)$, parameterized
by vector w, which is a good estimate of true probability p of membership of pattern
x in Class 1. This approach is a straightforward application to the principle explained in Sect. 2.1.2. The distribution concerned here is the Bernoulli distribution
with a single distribution parameter p.

For a pattern vector x_k and a scalar class indicator y_k, the likelihood of the probability p resulting as a function $f(x, w)$ is

$$f(x_k, w), y_k = 1$$
$$1 - f(x_k, w), y_k = 0 \tag{2.55}$$

This can be written more compactly as

$$f(x_k, w)^{y_k}(1 - f(x_k, w))^{1-y_k} \tag{2.56}$$

where the exponents y_k and $1 - y_k$ acquire values 0 or 1 and thus "select" the correct alternative from (2.55).

For a sample (or training set) of mutually independent samples, the likelihood over this sample is the product

$$\prod_{k=1}^{K} f(x_k, w)^{y_k} (1 - f(x_k, w))^{1-y_k}$$

$$= \prod_{k=1, y_k=1}^{K} (f(x_k, w)) \prod_{k=1, y_k=0}^{K} (1 - f(x_k, w)) \tag{2.57}$$

Maximizing (2.57) is the same as minimizing its negative logarithm

$$L = -\sum_{k=1}^{K} y_k \ln f(x_k, w) + (1 - y_k) \ln (1 - f(x_k, w))$$

$$= -\sum_{k=1, y_k=1}^{K} \ln (f(x_k, w)) - \sum_{k=1, y_k=0}^{K} \ln (1 - f(x_k, w)) \tag{2.58}$$

If the training set is a representative sample from the statistical population associated with pattern x_k, the expected value of likelihood per pattern L/K can be evaluated. The only random variable in (2.58) is the class indicator y, with probability p of being equal to one and $1 - p$ of being zero:

$$E[L/K] = E[y_k \ln (f(x_k, w)) + (1 - y_k) \ln (1 - f(x_k, w))]$$
$$= p(x_k) \ln (f(x_k, w)) + (1 - p(x_k)) \ln (1 - f(x_k, w)) \tag{2.59}$$

The minimum of this expectation is where its derivative with regard to the output of mapping $f()$ is zero:

$$\frac{\partial E[L]}{\partial f} = p(x_k) \frac{\partial}{\partial f} \ln (f(x_k, w)) + (1 - p(x_k)) \frac{\partial}{\partial f} \ln (1 - f(x_k, w))$$

$$= \frac{p(x_k)}{f(x_k, w)} - \frac{1 - p(x_k)}{1 - f(x_k, w)} = 0 \tag{2.60}$$

$$p(x_k)(1 - f(x_k, w)) - (1 - p(x_k)) f(x_k, w) = 0$$

$$f(x_k, w) = p(x_k)$$

This means that if the mapping $f()$ is expressive enough to be parameterized to hit the conditional class 1 probability for all input patterns x, this can be reached by minimizing the log-likelihood (2.58). In practice, a perfect fit will not be possible. In particular, with a mapping $f(x) = Bx$, it is clearly nearly impossible because of outputs Bx that would probably fail to remain in the interval $(0, 1)$. Also, with a logistic regression (2.52), it will only be an approximation for which no analytical solution is known. However, iterative numerical methods lead frequently to good results.

As an alternative to the maximum likelihood principle, a least squares solution minimizing the square deviation between the forecast and the true class indicator can be considered.

For a sample (x_k, y_k), the error is

$$e_k = (f(x_k, w) - y_k)^2 \tag{2.61}$$

The mean value of the error over the whole sample population, that is, MSE, is

$$\begin{aligned} E &= E\left[(f(x_k, w) - y_k)^2\right] \\ &= p(x_k)(f(x_k, w) - 1)^2 + (1 - p(x_k))(f(x_k, w))^2 \end{aligned} \tag{2.62}$$

This is minimum for values of $f()$ satisfying

$$\begin{aligned} \frac{\partial E}{\partial f} &= \frac{\partial}{\partial f}\left(p(x_k)(f(x_k, w) - 1)^2 + (1 - p(x_k))(f(x_k, w))^2\right) \\ &= 2p(x_k)(f(x_k, w) - 1) + 2(1 - p(x_k))f(x_k, w) \\ &= 2(f(x_k, w) - p(x_k)) \\ &= 0 \end{aligned} \tag{2.63}$$

or

$$f(x_k, w) = p(x_k) \tag{2.64}$$

Obviously, minimizing MSE is equivalent to the maximum likelihood approach, supposed the parameterized approximator $f(x, w)$ is powerful enough for capturing the dependence of class 1 probability on the input pattern vector.

Although the least square measure is not strictly identical with the misclassification loss, they reach their minimum for the same parameter set (assuming the sufficient representation power of the approximator, as stated above). In asymptotic terms, the least squares are close to zero if the misclassification loss is close to zero, that is, if the classes are separable. However, for strictly separable classes, there is a singularity—the optimum parameter set is not unambiguous and the parameter vector may grow without bounds.

With a parameterized approximator $f(x, w)$ that can exactly compute the class probability for a given pattern x and some parameter vector w, the exact fit is at both the maximum of likelihood and the MSE minimum (i.e., least squares). Of course, to reach this exact fit, an optimization algorithm that is capable of finding the optimum numerically has to be available. This may be difficult for strongly nonlinear approximators.

Least squares with logistic activation function seem to be the approach to classification that satisfies relatively well the requirements formulated at the beginning of Sect. 2.2.

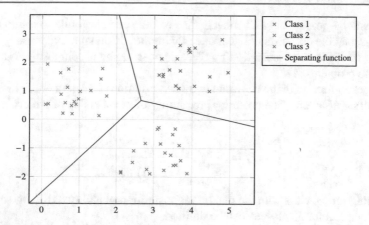

Fig. 2.16 Separation of three classes

2.2.4   Generalization to Multiple Classes

All classification principles of Sects. 2.2.1–2.2.3 can be generalized to multiple classes. Class separation is done by determining the maximum of the indicator functions. For linear indicator functions, this leads to linear separating lines such as shown in Fig. 2.16.

There is a straightforward generalization of the perceptron rule (2.28). Every class m has its own class indicator function $f_m(x, w)$. The forecast class assignment is

$$q_k = \arg\max_{m'} f_{m'}(x_k, w)$$

The output pattern y_k is a unit vector of true class indicators y_m with unity at the position of true class p_k and zeros otherwise.

Suppose the misclassification loss if class p is erroneously classified as q is L_{pq}. The error function is that of Hrycej [9]:

$$E = L_{pq} r \left(f_q(x_k, w) - f_p(x_k, w) \right)$$
$$r(z) = \begin{cases} z, & z > 0 \\ 0, & z \leq 0 \end{cases} \tag{2.65}$$

with partial derivatives

$$\frac{\partial E}{\partial f_q} = L_{pq} s \left(f_q(x_k, w) - f_p(x_k, w) \right)$$
$$\frac{\partial E}{\partial f_p} = -L_{pq} s \left(f_q(x_k, w) - f_p(x_k, w) \right)$$
$$\frac{\partial E}{\partial f_{q'}} = 0, \quad q' \neq p, q \tag{2.66}$$
$$s(z) = \begin{cases} 1, & z > 0 \\ 0, & z \leq 0 \end{cases}$$

It has to be pointed out that convergence of the perceptron rule to a stable state for non-separable classes is not guaranteed without additional provisions.

Also, multiple class separation by SVM is possible by decomposing to a set of two-class problems [4].

The generalization of the Bayesian two-class criterion (2.32) to a multiclass problem is straightforward. The posterior probability of membership of pattern x in the ith class is

$$P_i = \frac{f(x \mid i)\, p_i}{\sum_{j=1}^{M} f(x \mid j)\, p_j} \tag{2.67}$$

Seeking for the class with the maximum posterior probability (2.67), the denominator, identical for all classes, can be omitted.

Under the assumption that the patterns of every class follow multivariate normal distribution, the logarithm of the numerator of (2.67) is

$$\ln\left(\frac{1}{\sqrt{(2\pi)^N |C_i|}}\right) - \frac{1}{2}(x - m_i)'\, C_i^{-1}\,(x - m_i) + \ln(p_i)$$

$$= -\frac{N}{2}\ln(2\pi) - \frac{1}{2}\ln(|C_i|) - \frac{1}{2}x'C_i^{-1}x + m_i'C_i^{-1}x - \frac{1}{2}\left(m_i'C_i^{-1}m_i\right) + \ln(p_i) \tag{2.68}$$

which can be organized to the quadratic expression

$$q_i(x) = x'A_i x + b_i' x + d_i \tag{2.69}$$

with

$$A_i = -\frac{1}{2}C_i^{-1}$$

$$b_i' = m_i'C_i^{-1} \tag{2.70}$$

$$d_i = -\frac{N}{2}\ln(2\pi) - \frac{1}{2}\ln(|C_i|) - \frac{1}{2}\left(m_i'C_i^{-1}m_i\right) + \ln(p_i)$$

The Bayesian optimum which simultaneously minimized the misclassification rate is to assign pattern x_k to class i with the largest $q_i(x_k)$. Assuming identical covariance matrices C_i for all classes, the quadratic terms become identical (because of identical matrices A_i) and are irrelevant for the determination of misclassification rate minimum. Then, the separating functions become linear as in the case of two classes.

2.3 Dynamical Systems

Generic applications considered so far have a common property: an input pattern is mapped to an associated output pattern. Both types of patterns occur and can be observed simultaneously. This view is appropriate for many application problems such as processing static images or classification of credit risk. However, some application domains are inherently dynamic. In speech processing, semantic information is contained in the temporal sequence of spoken words. Economic forecasts are concerned with processes evolving with time—the effects of a monetary measure will occur in weeks, months, or years after the measure has been taken. Also, most physical and engineering processes such as wearing of technical devices or reaction of a car to a steering wheel movement evolve in time. Their output pattern depends not only on the current input pattern but also on past ones. Such systems are referred to as *dynamical systems*.

Quantitative relationships, including dynamical systems, are usually described with the help of equations. An equation can be viewed as a mapping of its right side to its left side (or inversely). If one of both sides consists of a single scalar or vector variable, it is possible to capture the description of such systems in a general mapping. In contrast to static relationships, input and output patterns in such mappings will encompass features or measurements with origin at varying times.

Let us consider the ways to describe a dynamic system by a mapping more concretely. One of the variants implies the assumption that the output depends only on past values of the input. For a scalar output y and scalar input x, both indexed by the discrete time point t, the linear form is

$$y_t = \sum_{h=0}^{H} b_h x_{t-h} \qquad (2.71)$$

The output is a weighted sum of past inputs, and it is independent of past inputs delayed by more than H time steps. In other words, the output response to an impulse in an input is terminated in a finite number of time steps. This is why this model is called *Finite Impulse Response* (FIR) in signal theory [14]. Its analogy in the domain of stochastic process models is the *moving average model* using white noise input [5].

To see which class of dynamical systems can be described by this mapping, it is necessary to review some basics of system theory. As in many other domains, its postulates are most profound for linear systems.

Every linear dynamical system can be, beside other description forms, completely specified by its *impulse response*. Since the available data typically consist of a set of measurements at discrete time points, we will consider discrete time systems. An impulse response is the trajectory of the system output if the input is a unit impulse at a particular time step, for example at $t = 0$ while it is zero at other steps. An example of a very simple system is shown in Fig. 2.17. It is the so called *first-order system*, an example of which is a heat exchange process. After heating one end of a metal rod to a certain temperature, the temperature of the other end will change proportionally to the temperature difference of both ends. To a temperature impulse at one end, the other end's temperature will follow the curve shown in Fig. 2.17.

More exactly, the response to a unit impulse is a sequence of numbers b_h, corresponding to the output h time steps after the impulse. Impulse response is a perfect characteristic of a linear system for the following reasons. Linear systems have the property that the response to a sum of multiple signals is equal to the sum of the responses to the individual signals. Every signal sequence can be expressed as a sum

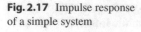

Fig. 2.17 Impulse response of a simple system

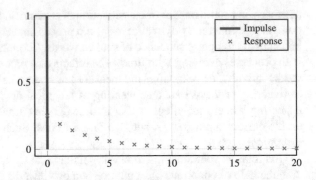

of scaled impulses. For example, if the signal consists of the sequence $(3, 2, 4)$ at times $t = 0, 1, 2$, it is equivalent to the sum of a triple impulse at time $t = 0$, a double impulse at $t = 1$, and a quadruple impulse at $t = 2$. So, the output at time t is the sum of correspondingly delayed impulse responses. This is exactly what the equality (2.71) expresses: the output y_t is the sum of impulse responses b_h to delayed impulses scaled by input values x_{t-h}, respectively.

The response of most dynamical systems is theoretically infinite. However, if these systems belong to the category of stable systems (i.e., those that do not diverge to infinity), the response will approach zero in practical terms after some finite time [1]. In the system shown in Fig. 2.17, the response is close to vanishing after about 15 time steps. Consequently, a mapping of the input pattern vector consisting of the input measurements with delays $h = 0, \ldots, 15$ may be a good approximation of the underlying system dynamics.

For more complex system behavior, the sequence length for a good approximation may grow. The impulse response of the system depicted in Fig. 2.18 is vanishing in practical terms only after about 50 time steps. This plot describes the behavior of a *second-order system*. An example of such system is a mass fixed on an elastic spring, an instance of which being the car and its suspension. Depending on the damping of the spring component, this system may oscillate or not. Strong damping as implemented by faultless car shock absorbers will prevent the oscillation while insufficient damping by damaged shock absorbers will not.

With further growing system complexity, for example, for systems oscillating with multiple different frequencies, the pattern size may easily grow to thousands of time steps. Although the number of necessary time steps depends on the length of the time step, too long time steps will lead to a loss of precision and at some limit value even cease to represent the system in an unambiguous way.

The finite response representation of dynamic systems is easy to generalize to nonlinear systems. The mapping sought is

$$y_t = f\left([x_t \ldots x_{t-h}]', w\right) \tag{2.72}$$

for training samples with different indices t, with correspondingly delayed input patterns. A generalization to multivariate systems (with vector output y_t and a set of delayed input vectors x_{t-h}) is equally easy. However, the length of a necessary

Fig. 2.18 Impulse response of an oscillating system

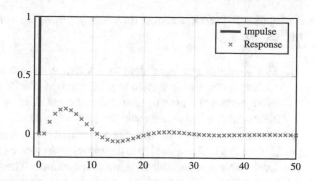

response sequence is a serious constraint for applicability of finite response representation of dynamic systems.

Fortunately, system theory provides alternative representation forms. A model of the general form (mathematically, a *difference equation*)

$$x_{t+1} = Ax_t + Bu_t \tag{2.73}$$

is called a *discrete linear system in state-space form*, with t being an index running with time. The notation usual in system theory is different from that used otherwise in this book. The vector variable u_k denotes an external input (or *excitation*) to the system. The vector x_k represents the state of the system (not the input, as in other sections of this work), obviously subject to feedback via matrix A.

Note that in the neural network community, the system input is frequently denoted by x. Here, the labeling common in systems theory is used: x for state and u for input. In some parts of this book (e.g., Sect. 6.4), the neural network community denotation is adopted.

In some systems, the state variables are not observable. Then, observable variables y are introduced, resulting from the state and input variables in an algebraic (non-dynamic) manner:

$$y_t = Cx_t + Du_t \tag{2.74}$$

In our computing framework, all state variables can be accessed, so that (2.73) is sufficient.

Time flows in discrete steps. In physical or economical systems, the steps are typically equidistant and corresponding to some time period such as a second, a millisecond, or a year, depending on the application. The term *discrete* lets us distinguish models running in such discrete time periods in contrast to *continuous* time models described by differential equations instead of difference equations. In settings such as *Recurrent Neural Network* (RNN) discussed in Sect. 3.3, the index k is simply assigned to computing cycles for the evaluation of a neural network, no matter what duration these cycles have.

Some aspects of behavior of linear systems (also as approximations of nonlinear systems) are important for understanding DS models and will now be briefly explained. First of all, it is clear that the response of such system is potentially infinite.

An input will result in a certain trajectory in time. There are several variants of these trajectories:

1. The activations of individual nodes can converge to a certain terminal state. Although this terminal state will be reached only after infinite time, the values will approach this state by some rate inherent to the totality of system parameters. This behavior is characteristic for *stable systems*.
2. In other parametric variants, the system can be *unstable*. In this case, the activation values will diverge from the initial state toward infinity.
3. Both stable and unstable systems can *oscillate*. Then, the values are fluctuating up and down with some characteristic frequency or a mixture of such frequencies. The oscillating amplitude of stable systems will decrease toward zero (in infinite time) while that of unstable systems will infinitely increase.

It is particularly the instability which may lead to a complete failure of the intended application, including numerical overflow.

These behavioral alternatives result from the state matrix A, or, more exactly, from its *eigenvalues*. Eigenvalues λ are an important concept of linear algebra. They are complex-valued solutions of equation

$$Ax = \lambda x \tag{2.75}$$

for a fixed square matrix A and some unknown vector x and can easily be computed by most numerical packages. A matrix with n dimensions has n eigenvalues. In systems and control theory, the eigenvalues of matrix A are referred to as *poles* of the system.

Stable systems are those where all eigenvalues have an absolute value not greater than unity, that is, $|\lambda| \leq 1$. Any eigenvalue greater than unity indicates instability. This can be intuitively understood by observing (2.73). Multiplying with A will, for states x from (2.75), result in diminishing norms of the states if $|\lambda| \leq 1$ and growing norms if $|\lambda| > 1$.

Oscillating behavior is a property of systems with complex, rather than real, eigenvalues. The size of the imaginary part determines the oscillating frequency.

The representation by a state-space model is equally appropriate for univariate and multivariate inputs and outputs. On the other hand, it is not easy to identify such models from data. For linear models, the subspace method from Van Overschee and De Moor [20] receives a state-space model by an algebraic transformation of input and output series. Its generalization to nonlinear models is not straightforward. An outstanding problem remains the guarantee of model stability. Most linear systems are stable—otherwise, they would inevitably converge to infinite state and output values. But it is not guaranteed that a model received from data measured on a stable linear system shares the stability property. From the viewpoint of DS, the subspace method is fundamentally different than fitting a parameterized mapping to input and output pairs. The reason is the internal feedback of non-observable state variables. To summarize, using state-space models in typical contexts of DS where model identification from data is a key part of the task is difficult and may require expert knowledge and experience in systems theory.

Fortunately, alternatives exists. In addition to the state-space representation (2.73) and (2.74) of a linear dynamical system, there is an equivalent representation by *transfer functions*. The equivalence of both system descriptions (state space and transfer function) is a fundamental finding of linear systems theory. Its detailed justification would go beyond the scope of this book.

The transfer function representation takes place in terms of delayed values of input and output. A transfer function with a scalar input and a scalar output is a recursive linear equation:

$$y_t = \sum_{i=1}^{m} a_i y_{t-i} + \sum_{i=0}^{n} b_i u_{t-i} \tag{2.76}$$

This process description inherently captures an Infinite Impulse Response and is thus called *Infinite Impulse Response* (IIR) filter in signal theory.

Statisticians may notice that it is closely related to the *Autoregressive Moving Average* (ARMA) model used in statistical modeling, where input u is usually a random disturbance. The sum of feedforward inputs with coefficients b is the moving average part, the feedback with coefficients a being the autoregressive one.

Compared to the FIR representation, transfer functions are substantially more economical in the number of parameters. The system with impulse response of Fig. 2.17 is completely specified by two parameters: $a_1 = 0.7487$ and $b_0 = 0.2513$. The description of the oscillating system of Fig. 2.18 is specified by the tree parameters: $a_1 = 1.68377$, $a_2 = -0.78377$, and $b_0 = 0.10000$.

The generalization to a nonlinear system with a scalar input and scalar output is straightforward:

$$y_t = f\left(\left[y_{t-1} \cdots y_{t-m} \, u_t \cdots u_{t-n}\right]', w\right) \tag{2.77}$$

For both linear and nonlinear mapping, the training set has a usual form of input/output pairs

$$\left(\left[y_{t-1} \cdots y_{t-m} \, u_t \cdots u_{t-n}\right]', y_t\right) \tag{2.78}$$

that can be fitted by the mean square method.

It is an important decision how many delayed input and output variables are to be used. It has to do with the *order* of the modeled system. In the linear state-space representation (2.73) and (2.74), the order is equal to the size of square matrix A, no matter whether the input u and the output y are scalars or vectors. Every scalar linear model (called *Single Input/Single Output* (SISO)) in the state-space form can be transformed to an equivalent model in the transfer function form (2.76) with $n \leq m$. The system poles, i.e., eigenvalues λ satisfying (2.75), are simultaneously the roots of the polynomial

$$z^m - \sum_{i=1}^{m} a_i z^{m-1} = 0 \tag{2.79}$$

where the operator z corresponds to one discrete step ahead in time. This is based on rewriting (2.76) as

$$y_t - \sum_{i=1}^{m} a_i y_{t-i} = \sum_{i=0}^{n} b_i u_{t-i} \qquad (2.80)$$

and substituting powers of z for variable delays, resulting in the *frequency domain representation* of the transfer function:

$$\left(z^m - \sum_{i=1}^{m} a_i z^{m-i} \right) y = \left(\sum_{i=0}^{n} b_i z^{m-i} \right) u$$

or

$$y = \frac{\sum_{i=0}^{n} b_i z^{m-i}}{z^m - \sum_{i=1}^{m} a_i z^{m-i}} u \qquad (2.81)$$

The transformation of a state-space model with vector input and output (*Multiple Input/Multiple Output* (MIMO)) to the transfer function form is more complex. The result is a matrix of transfer functions from each input variable to each output variable. All these transfer functions have poles from the same pole set (corresponding to the eigenvalues of matrix A in the corresponding state-space representation) but each single transfer function may have only a small subset of them. Consequently, the order of each single denominator polynomial is a priori unknown. It is this order which determines how many past values of input and output variables are to be considered to receive the correct model. This number is called *observability index* [11]. Its founded assessment for an unknown system, probably a typical situation in DS, is difficult. On the other hand, the identification of systems with observability indices of more than four is numerically risky because of measurement noise. So, a choice of three or four will mostly be appropriate. Then, the training set consists of data pairs (2.78) with delayed vectors u and y.

For general parameterized mapping types, training sets of this structure can be used for fitting the parameter set w, so that the mapping outputs (2.77) are close to the measured ones in the sense of MSE. Hence, widespread training algorithms for MSE can be used.

Unfortunately, there is no guarantee that the model received preserves the stability property. Although measured data may arise from stable system, the identified model is divergent. This instability is likely to lead to unusable results of even numeric overflows. A means for preventing this is to use longer measurement sequences as training examples and to make recursive simulation of the model mapping (2.77). The recursive simulation is implemented by feeding the simulated, instead of measured, delayed outputs as training example input. In (2.77), it would amount to taking simulated values of $y_{t-1} \ldots y_{t-m}$ received as output of previous training examples. So y_{t-1} would be received as output of the previous training example

$$y_{t-1} = f \left(\left[y_{t-2} \cdots y_{t-m-1} \, u_{t-1} \cdots u_{t-n-1} \right]', w \right) \qquad (2.82)$$

The mapping $f()$ being unstable would manifest itself in growing MSE. So minimizing such multistep error measure will, with a sufficiently long training sequence, prevent the instability [10].

To summarize the findings about representation of dynamical systems:

> Dynamical systems can be represented either in the strictly feedforward FIR form (2.72) or in the IIR form (2.77) containing outer feedback.
>
> The FIR form may require a long input history to be a good approximation. The absence of feedback makes it insensitive to stability issues.
>
> The IIR form is substantially more compact. Its inherently dynamical character implies instability potential. Stability problems may require using a multistep error measure with corresponding training sequences.

2.4 Spatial Systems

The overview of important mappings that can be modeled and fitted to data would not be complete without briefly mentioning spatial systems.

In natural sciences, dynamical systems of Sect. 2.3 are mostly described by differential equations, whose discrete time counterpart are the difference equations of type (2.72) or (2.77). For example, the discrete time system producing the impulse response of Fig. 2.17 can also be described in continuous time by the differential equation

$$\frac{\mathrm{d}y}{\mathrm{d}t} = -ay + bu \tag{2.83}$$

Differential equations containing only derivatives with respect to time t are called *ordinary differential equations*. Their generalization are *partial differential equations* containing derivatives with respect to multiple variables. These variables can be arbitrary domain-specific characteristics. Frequently, they are related to spatial positions in one or more dimensions. A simple example is the form of an elastic rod fixed horizontally at one its end on a wall. If its thickness is small relative to its length, it can be viewed as a one-dimensional line. Because of its elasticity, it will not retain the form of a perfect horizontal line. Rather, its free end will bend downwards by gravity. Starting in a horizontal position, its form will develop in time to finally assume a form in which the forces are in a static balance, as shown in Fig. 2.19.

Fig. 2.19 Development of the form of an elastic rod

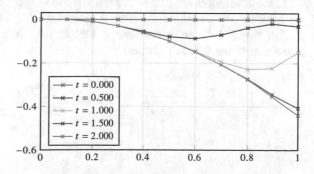

This behavior is described by the following partial differential equation:

$$r\left(\frac{\partial^2 y}{\partial t^2} - g\right) + d\frac{\partial y}{\partial t} = e\frac{\partial^2 y}{\partial x^2} \tag{2.84}$$

The variables x and y correspond to the horizontal and vertical positions of infinitely small rod elements: the element with position x from the wall fixing point has the horizontal position y. The equation expresses the dynamic balance of forces. At the left side, there are forces related to the vertical movement. First, it is the vertical acceleration (that would be equal to the earth acceleration g in absence of other forces) as a second derivative of the vertical position y with regard to time t. The force corresponding to this acceleration results by multiplication by the material-specific density r. Second, there is a material damping with damping coefficient d, proportional to the element's velocity (the first derivative of the position with regard to time t). At the right side, there is the elastic force acting against gravity. This elastic force (with material-specific elastic modulus e) is proportional to the local bend of the rod, characterized by the second derivative of the vertical position y with regard to the horizontal position x.

Data measurements will always be done at discrete times and discrete spatial measurement points. The adequate description means for this configuration is by difference equations, using discrete approximations of derivatives in which the infinitesimally small step dx in variable x is substituted by the discrete step Δx. The first partial derivative $\frac{\partial y}{\partial x}$ corresponds to

$$\frac{y_{t-1,i} - y_{t-2,i}}{\Delta x}. \tag{2.85}$$

The discrete analogy of the second partial derivative $\frac{\partial^2 y}{\partial x^2}$ is

$$\frac{y_{t,i} - 2y_{t-1,i} + y_{t-2,i}}{(\Delta t)^2}. \tag{2.86}$$

The difference equation counterpart of (2.84) is

$$r\left(\frac{y_{t,i} - 2y_{t-1,i} + y_{t-2,i}}{(\Delta t)^2} - g\right) + d\frac{y_{t-1,i} - y_{t-2,i}}{\Delta t}$$
$$= e\frac{y_{t-1,i} - 2y_{t-1,i-1} + y_{t-1,i-2}}{(\Delta x)^2} \tag{2.87}$$

Some differences are shifted one time step to the past to provide the possibility to compute the future values of y from the past ones.

Equation (2.87) can be rewritten so that the new value $y_{t,i}$ is at the left side and remaining terms at the right one:

$$y_{t,i} = 2y_{t-1,i} - y_{t-2,i}$$
$$+ \left(\frac{e}{r}\frac{y_{t-1,i} - 2y_{t-1,i-1} + y_{t-1,i-2}}{(\Delta x)^2} - \frac{d}{r}\frac{y_{t-1,i} - y_{t-2,i}}{\Delta t} + g\right)(\Delta t)^2 \tag{2.88}$$

Such equations can be expressed in the general mapping form, parameterized by vector w:

$$y_{t,i} = f\left(\left[y_{t-1,i}\ y_{t-1,i-1}\ y_{t-1,i-2}\ y_{t-2,i}\right], w\right) \tag{2.89}$$

Simulating a system by the application of equations such as (2.89) is called the *finite difference method* [18]. In Fig. 2.19, a selection of data points $y_{t,i}$ is symbolized by crosses at individual time points. (In the simulation, the time step has been 0.001 s.)

This representation of the physical system makes clear that the output value at time t and position i depends not only on the corresponding value at previous time points (here: $t-1$ and $t-2$) but also on its spatial neighbors (here: at positions $i-1$ and $i-2$). This is an extension against the pure dynamic systems.

This task can also be reformulated in the sense of DS. What is given are training examples as measured input/output pairs with index t varying along the time axis and index i varying along the positions on the rod.

$$\vdots$$

$$\left(\left[y_{t-1,i-1}\ y_{t-1,i-2}\ y_{t-1,i-3}\ y_{t-2,i-1}\right], y_{t,i-1}\right)$$
$$\left(\left[y_{t-1,i}\ y_{t-1,i-1}\ y_{t-1,i-2}\ y_{t-2,i}\right], y_{t,i}\right)$$
$$\left(\left[y_{t-1,i+1}\ y_{t-1,i}\ y_{t-1,i-1}\ y_{t-2,i+1}\right], y_{t,i+1}\right)$$

$$\vdots \tag{2.90}$$

$$\left(\left[y_{t,i-1}\ y_{t,i-2}\ y_{t,i-3}\ y_{t-1,i-1}\right], y_{t+1,i-1}\right)$$
$$\left(\left[y_{t,i}\ y_{t,i-1}\ y_{t,i-2}\ y_{t-1,i}\right], y_{t+1,i}\right)$$
$$\left(\left[y_{t,i+1}\ y_{t,i}\ y_{t,i-1}\ y_{t-1,i+1}\right], y_{t+1,i+1}\right)$$

$$\vdots$$

The stability issues of models of type (2.89) are even more complex than for dynamical systems and are to be considered [6]. Their discussion is beyond the scope of this book.

A related topic are spatial operators used in image processing. Analogically to the variables in the training set (2.90), they are functions of the neighboring variables in topologically ordered data sets. They are discussed in the context of neural networks in Sect. 3.6. In contrast to the spatial systems for which the output measurement is known, these operators are mostly fitted implicitly to be indirectly optimal for a high-level task such as image classification.

2.5   Mappings Received by "Unsupervised Learning"

The mappings considered so far have been fitted to a training set consisting of pattern pairs. Every pattern pair is constituted by an input pattern and an output pattern. The mapping fitted to these pairs implements an unidirectional association between both patterns: To every input pattern, an output pattern is assigned. This principle is frequently referred to as *Supervised Learning*.

This view is appropriate for most but not all DS tasks. Sometimes, the goal is not to find an association between input and output patterns but rather an advantageous representation of input patterns without specifying the desired output. "Advantageous" can comprise properties such as lower pattern dimension or mutual independence of pattern features.

This approach is usually motivated by impossibility to receive output patterns, or in the case of a classification task, by missing class labels (which constitute the output pattern in classification problems). This is, in turn, a consequence of insurmountable obstacles or a too large expense for labeling, in particular in tasks of considerable size such as millions of training examples. Two examples show typical difficulties with the labeling:

- In image processing, the labels consist in specifying the object depicted in the image.
- In corpus-based semantics, the labels may correspond to some form of specification of semantic meaning.

In both cases, the labeling is basically manual work of considerable extent. In the case of semantics, it is a task for highly skilled professionals.

This is why the tasks are reformulated in a way for which the availability of output pattern or labels is not strictly required. The approach for processing data in this way is usually called *Unsupervised Learning*. This term is somewhat misleading since there is no lack of supervision in the training procedures. The substance of the concept is only in the absence of explicit input/output pairs to be mimicked by the mapping sought.

Beyond missing output patterns, there are some further justifications of such approach:

- The favorable representation may be a basis for multiple applications.
- A supervised learning task can be excessively difficult from the numerical point of view. This is frequently the case for deep neural networks with many hidden layers. Fitting the weights becomes increasingly difficult with the growing distance from the output layer (e.g., because of vanishing gradient).
- The gained representation can be useful itself, for example, for recognizing the structure of the task. Such discovered structure can be a basis for decomposing the task into subtasks.
- Assumed associations are only implicit within the input pattern, for example, between consecutive words in a text.

The definition of the unsupervised concept by availability of output patterns is not the only one used. The algorithmic view may be different. The fact that output patterns are omitted does not automatically require using algorithms specific for unsupervised learning. In some cases, parts of the input pattern are extracted and used as output patterns. Then, supervised learning algorithms for fitting a mapping to input/output pairs can be used without change. In such cases, it is not strictly justified to speak about unsupervised learning. Nevertheless, some authors use the term in this extended way.

In the following, some mappings gained by the unsupervised paradigm are discussed.

2.5.1   Representations with Reduced Dimensionality

Objects or patterns can be described by feature vectors. Collecting all features that may be relevant for some DS task will probably result in a vector of a high dimension. In practice, it can be expected that some or even many features contain similar descriptive information. Which information is relevant for a task can be ultimately decided only after the task has been solved. However, some redundancies can also be discovered in general statistics terms. For example, if a feature is a function of another one, one of them is redundant. Even if such functional dependence is only approximate, there may still be a substantial extent of redundancy. Although discarding any features that are not exactly functionally dependent on others will always bring about some information loss, this loss may be justified if a considerable dimension reduction is the reward, making the task better tractable by numerical methods.

In linear terms, the amount of dependence between individual features is quantified by the covariance matrix. If feature variables are to be scaled proportionally to their variability, the correlation matrix is a more appropriate characteristic.

The standard linear dimensionality reduction method is the *Principal Component Analysis* (PCA), proposed by Pearson [15]. Its idea is to decompose the variability of a vector of numerical features to independent components of decreasing importance. The importance of the components is evaluated by their individual variability, following the principle that components with relatively small variability are less informative for most tasks. The dimension of the feature vector can then be reduced by selection components responsible for the bulk of total variability (measured as statistical variance), omitting less important ones.

Two dimensional patterns shown in Fig. 2.20 exhibit the largest variability along the axis $PC1$, the largest principal component, called *dominant* component. The component orthogonal to $PC1$ is $PC2$, capturing the remaining variability. The second principal component ($PC2$) is still considerably varying so that reducing the data to $PC2$ would bring about a substantial loss of information. In contrast, the patterns of Fig. 2.21 can be well characterized by their positions along component $PC1$ alone.

These two examples show that depending on data, a reduced representation can be found. This reduced representation sometimes retains most information, sometimes not.

Fig. 2.20 Patterns in $2D$
space, orthogonal
components of variability

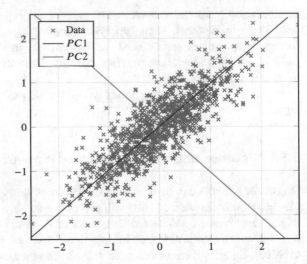

Fig. 2.21 Patterns in $2D$
space with a clearly
dominant variability
component

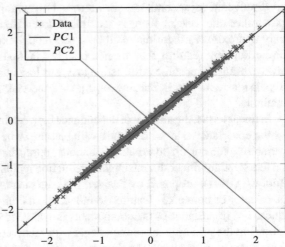

Mathematically, PCA is based on an important concept of linear algebra, the *eigenvalues* and *eigenvectors*. They were already mentioned in relationship with dynamical systems in Sect. 2.3. Eigenvalues of a square matrix A are solutions λ of the equation

$$|A - \lambda_i I| = 0 \tag{2.91}$$

Since the determinant in (2.91) is, for a (N, N) matrix A, a polynomial of Nth order, the equation has N solutions λ_i. For each eigenvalue, there is a right unit column vector v_i and a left unit column vector w_i satisfying

$$
\begin{aligned}
(A - \lambda_i I)\, v_i &= 0 \\
w_i'\, (A - \lambda_i I) &= 0
\end{aligned}
\tag{2.92}
$$

Numerical procedures for finding eigenvalues and eigenvectors are built-in functions in many mathematical packages.

Transforming (2.92) to

$$Av_i = \lambda_i v_i$$
$$w_i' A = w_i' \lambda_i \tag{2.93}$$

makes obvious that multiplying the eigenvector by the corresponding eigenvalue is equivalent to the multiplication by matrix A.

In contrast to characterizing the behavior of dynamic systems, where matrix A was square but arbitrary, we are now particularly interested in properties of eigenvectors and eigenvalues of symmetric matrices.

For a symmetric matrix A, both eigenvectors are obviously identical. They can be combined in the equation

$$v_i' A v_i = \lambda_i v_i' v_i = \lambda_i \tag{2.94}$$

For all N eigenvectors, (2.94) can be organized into a matrix form

$$V' A V = L \tag{2.95}$$

with a diagonal matrix L with elements λ_i.

The additional property with a symmetric A is that the eigenvectors are orthonormal so that $VV' = I$ and $V' = V^{-1}$.

These are the key properties for PCA. The covariance (or correlation in the sense of the remark above) matrix C of the input pattern vector x is by its nature symmetric. Each eigenvector v_i defines the transformation of x to the ith principal component $v_i x$. The variance of this component is, according to (2.94),

$$v_i' C v_i = \lambda_i \tag{2.96}$$

The complete covariance matrix of the principal components is diagonal

$$V' C V = L \tag{2.97}$$

that is, the components are independent with zero covariance.

If the eigenvectors are ordered, the largest variance is attributed ot the first component corresponding to the largest eigenvalue λ_1 ($PC1$ in Figs. 2.20 and 2.21). M principal components with M largest eigenvalues have the proportion

$$\frac{\sum_{i=1}^{M} \lambda_i}{\sum_{i=1}^{N} \lambda_i} \tag{2.98}$$

of total variance $\sum_{i=1}^{N} \lambda_i$. This makes a reduced pattern made of M principal components the best linear approximation of size M. The input/output mapping in the sense of unsupervised learning is represented by

$$y = f(x) = V_M' x \tag{2.99}$$

with matrix V_M consisting of M columns of V with the largest eigenvalues.

The task of finding the largest principal component can also be formulated as an optimization task. This maximum provides valuable insight for DS applications solved by numerical optimization algorithms using gradient. The term to be maximized is the component variance

$$H = v_1' C v_1 \tag{2.100}$$

with an additional condition of unit vector length:

$$v_1' v_1 = 1 \tag{2.101}$$

With input pattern vectors x shifted to zero mean, the sample covariance matrix of the training set X is

$$C = \frac{1}{K} X' X = \frac{1}{K} \sum_{k=1}^{K} x_k x_k' \tag{2.102}$$

The term (2.100) to be maximized becomes

$$H = \frac{1}{K} \sum_{k=1}^{K} v_1' x_k x_k' v_1 \tag{2.103}$$

Its gradient with respect to v_1 is

$$\frac{\partial H}{\partial v_1} = \frac{2}{K} \sum_{k=1}^{K} x_k x_k' v_1 \tag{2.104}$$

For each pattern x_k, it is the product of the pattern and the scalar $x_k' v_1$. This scalar can be interpreted as a similarity measure between pattern x_k and the weight vector v_1. So, the gradient points to the direction of maximum similarity of the patterns and the weight vector.

Of course, the normalizing constraint (2.101) cannot be omitted. It can be approximated by including a penalty term

$$-c \left(v_1' v_1 - 1 \right)^2 \tag{2.105}$$

with appropriate constant c and gradient

$$-2c \left(v_1' v_1 - 1 \right) v_1 \tag{2.106}$$

Adapting the weights in the gradient direction thus corresponds to an *unsupervised learning rule*.

It has to be pointed out that this is a maximizing task, in contrast to minimizing tasks treated in other sections. This is why the learning rule is ascending (in the gradient direction) rather than descending (against the gradient direction).

A generalization of this rule to the space spanned by multiple principal components, that is, an optimal dimensionality reduction to M features, is given by Hrycej [8].

These learning rules are interesting in incremental learning context and as insights to possible biological learning mechanism. However, for information processing in the sense of Data Science, linear algorithms based on eigenvectors and eigenvalues are doubtlessly more efficient.

Another important insight has to do with the fact that the compressed feature vector resulting from M largest eigenvalues maximizes the proportion of the total variance (2.98). A direct consequence of this is that this feature vector is loaded with inevitable difference to the original vector. The square deviation corresponds to remaining $N - M$ eigenvalues. This property is discussed and exploited in the following section.

2.5.2 Optimal Encoding

Reducing dimensionality consists of determining a mapping of an input pattern to a reduced dimensionality pattern as in (2.99). A closely related task is determining an encoding such that the original input pattern can be restored with the least error. Formally, a pair of nested mappings $f(x)$ and $g(x)$ are sought

$$z = g(y) = g(f(x)) \tag{2.107}$$

where z is as close to x as possible. Mapping $f(x)$ is the encoding function while $g(x)$ is the decoding one. The quadratic error measure (corresponding to the MSE of z) to be minimized over a training set (consisting of input patterns x_k alone) is

$$E = \frac{1}{K} \sum_{n=1}^{N} \sum_{k=1}^{K} (z_{kn} - x_{kn})^2 \tag{2.108}$$

It corresponds to the trace (the sum of the diagonal elements of the matrix)

$$E = \frac{1}{K} \sum_{k=1}^{K} (z_k - x_k)(z_k - x_k)' \tag{2.109}$$

The special case of linear mapping can be treated analytically. The elements of encoding vector y_m can be set equal to principal components. The mappings for the mth component alone are

$$y_m = v'_m x$$
$$z = v_m y_m \tag{2.110}$$

This former mapping determines the component of vector x projected to the eigenvector v_m, the latter the projection back to the space of vector x. For M principal components and a single input pattern vector x, this can be expressed in matrix form

$$y = f(x) = V'_M x$$
$$z = g(y) = V_M y \tag{2.111}$$

that is,

$$z = g(f(x)) = V_M V'_M x \tag{2.112}$$

The vector deviation is

$$z - x = V_M V'_M x - x = \left(V_M V'_M - I \right) x \tag{2.113}$$

and the square error matrix

$$
\begin{aligned}
e &= (z - x)(z - x)' \\
&= x' \left(V_M V_M' - I\right)' \left(V_M V_M' - I\right) x \\
&= x' \left(V_M V_M' V_M V_M' - 2 V_M V_M' + I\right) x \\
&= x' \left(I - V_M V_M'\right) x \\
&= x'x - x'V V'x - x' V_M V_M' x
\end{aligned}
\tag{2.114}
$$

The average of (2.114) over the training set as in (2.109) is

$$
\begin{aligned}
E &= \frac{1}{K} \sum_{k=1}^{K} (z_k - x_k)(z_k - x_k)' \\
&= \frac{1}{K} \sum_{k=1}^{K} x_k' V V' x_k - \frac{1}{K} \sum_{k=1}^{K} x_k' V_M V_M' x_k
\end{aligned}
\tag{2.115}
$$

The first term corresponds to the sum of variances of all principal components $y_{ki} = v_i x_k, i = 1, \ldots, N$ of the sample covariance matrix over the training set while the second to the sum of the M largest ones: $y_{ki} = v_i x_k, i = 1, \ldots, M$. According to (2.96), the variances of principal components are equal to the corresponding eigenvalues λ_i. So, (2.115) equals to the sum of $N - M$ smallest eigenvalues of the covariance matrix:

$$
\sum_{i=M+1}^{N} \lambda_i
\tag{2.116}
$$

This is the minimum possible error of the approximation of input patterns x by patterns y of lower dimension. It is an important finding that optimum encoding can be found by least squares in the linear case. It motivates the use of least squares also for nonlinear mappings. While the minimum error of linear encoding problem possesses this closed form based on eigenvalues and eigenvectors, general nonlinear encoding can be sought by numerical minimization.

The encoding and decoding mappings are parameterized by the vector $w = \left[w_f \ w_g\right]$:

$$
z = g\left(y, w_g\right) = g\left(f(x, w_f), w_g\right)
\tag{2.117}
$$

The error function (2.108) is then minimized by some method using its gradient.

Formulation of application tasks as encoding problems is a popular and fruitful approach wherever a compact representation of large feature spaces is important. A prominent domain is corpus-based semantics where finding a compact semantic representation of words and phrases is the key challenge. Sophisticated variants may use modified decoding targets different from the complete input pattern. For example, masking parts of input and using the masked terms as decoding target is the principle behind *Bidirectional Encoder Representations from Transformers* (BERT) [3].

2.5.3   Clusters as Unsupervised Classes

Mappings for dimensionality reduction (Sect. 2.5.1) and optimal encoding (Sect. 2.5.2) are inherently continuous. A continuous (i.e., real valued) input vector is mapped to an equally continuous output vector of reduced dimensionality.

In a direct analogy to classification tasks discussed in Sect. 2.2 where a continuous input pattern vector is mapped to an assignment to one of discrete classes, an unsupervised version of classification is conceivable. In this case, discrete classes are not predefined but rather sought in the data. Such "natural" classes are called *clusters*.

Seeking clusters not defined in advance presumes some concept of evaluation how good a partitioning is. As in other tasks of unsupervised learning, there is no unambiguous answer to this. The variety of approaches is substantially larger than dimensionality reduction or optimal encoding. The only consensus is that objects within every cluster are to be more similar to each other than objects from different clusters. And it is already this similarity that can be defined in different ways.

One possibility is the similarity defined by variance. Let us consider objects described by a single scalar feature x (i.e., whose pattern vector is scalar) and their given partitioning to L clusters of K_l objects with $\sum_{l=1}^{L} K_l = K$. Their total variance can be decomposed in the following way:

$$
\begin{aligned}
V &= \frac{1}{K} \sum_{l=1}^{L} \sum_{k=1}^{K_l} (x_{lk} - \overline{x})^2 \\
&= \frac{K_l}{K} \sum_{l=1}^{L} \frac{1}{K_l} \sum_{k=1}^{K_l} (x_{lk} - \overline{x_k})^2 + \sum_{l=1}^{L} \frac{K_l}{K} (\overline{x_l} - \overline{x})^2 \\
&= V_{\text{in}} + V_{\text{cl}}
\end{aligned}
\tag{2.118}
$$

The first decomposition term V_{in} is the mean intracluster variance, or, more exactly, the weighted mean of variances of individual clusters. The second term, V_{cl}, is the weighted variance of cluster means or intercluster variance. The smaller the proportion of intracluster variance, the better the clustering.

At first glance, this decomposition may seem to be a good starting point for meaningful clustering. However, with a pattern vector of dimension beyond one, first problems become obvious. Individual pattern features have to be summarized in some way. A straightforward and frequently followed way would be to add the variances of individual features. The two-dimensional pattern set of Fig. 2.22 suggests an intuitive partitioning into clusters given in Fig. 2.23. However, the variance decomposition suggests only separation between clusters 1 and 2 along feature X_1 while feature X_2 separates only cluster 3 from the totality of clusters 1 and 2. This makes obvious that this straightforward extension of (2.118) may lead to undesired results.

Another problem of this criterion is that it can hardly be a complete basis for a practical clustering algorithm. The decomposition can be only be made for a given partitioning into clusters. There is a huge number of such decompositions equal to

$$
K^L
\tag{2.119}
$$

Fig. 2.22 Two-dimensional
patterns

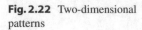

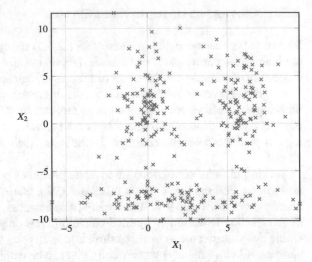

Fig. 2.23 Two-dimensional
patterns with suggested
clusters

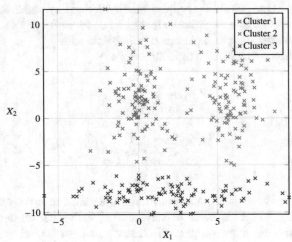

resulting from the possibility to assign every from K objects to any of L clusters.
Beyond this, the number of desired clusters is usually a priori unspecified, since the
goal is rather to have distinct clusters than a certain number thereof.

A partial remedy of the complexity problem is the following iterative procedure:

1. starting with a set of initial clusters;
2. computing the mean $\overline{x_l}$ of each cluster;
3. reassigning every object to the cluster with the smallest distance $\|x_{lk} - \overline{x_k}\|$;
4. recomputing the means from step 2.

Obviously, reassigning an object to the cluster with the nearest mean will reduce the intracluster variance (2.118). This is strictly true if the distance measure is

Euclidean, that is, a square norm. This algorithm known as the *k-means clustering algorithm* [13] converges to a state in which no reassignment is necessary.

Unfortunately, the cluster assignment in which the algorithm terminates is only a local minimum. Other assignments, possibly more advantageous with respect to (2.118), may exist. Since the local minimum to which the algorithm converges depends on the initial cluster assignment, it is usual to start from several such initial cluster sets and select the best of the local minima. Such initial sets can be defined either by their initial cluster means and object assignment according to the nearest mean, or by another heuristic clustering method. There is a plenty of such heuristic methods and their enumeration would go beyond the scope of this chapter. It is difficult to select the optimal, or even a reasonably good one for an application. The results are varying with properties and size of the data set.

Clustering of data patterns is an interesting possibility to introduce structure into an otherwise nontransparent data set. Clustering methods have their place in the stage of data exploration.

On the other hand, the clustering task is scarcely defined with mathematical rigor and the algorithms do not guarantee a globally optimal solution. Ultimately, a method with which the user is familiar and whose efficient implementation is available is to be preferred.

2.6 Chapter Summary

Most DS tasks consist of finding a mapping between an input pattern and an output pattern, expressing the association or relationship between both patterns. This mapping is to be consistent with the given measured data (consisting of such pattern pairs). The quality of the fit is measured by an error function reflecting the priorities of the task.

For linear mappings, optimum fit is frequently known in an analytical form. By contrast, fitting nonlinear mappings such as neural networks can mostly be done only by numerical optimization methods. This is why it is desirable that this error function can be directly used with such numerical methods.

- For continuous mappings, the error function (typically the MSE) can be directly minimized by an analytical expression or numerical algorithm.
- For classification mappings, the application-oriented error function (e.g., the misclassification rate) is frequently difficult to minimize directly. So, the function genuinely minimized is a different one, which may lead to contradictions. There is a variety of formulations trying to abridge this gap.
- Dynamic and spatial systems can also be formulated in terms of input/output mapping. However, they have characteristic pitfalls: they are potentially unstable or

oscillating. To avoid such behavior requires additional, sometimes tedious, checks of results delivered by the fitting procedure.

- Unsupervised learning is a framework for the treatment of data without explicit output patterns. What is sought is a mapping of measured input patterns to non-measured output patterns with some advantageous properties, e.g., reduced dimensionality.

2.7 Comprehension Check

1. What is the most frequently used error function for continuous mappings?
2. Which property of misclassification loss makes it inappropriate for numerical minimization?
3. Which output nonlinearity (i.e., nonlinear activation function of the output layer) is advantageous for probabilistic classification?
4. Which error functions (to be minimized) are adequate for probabilistic classification?
5. Dynamic systems can be represented by a mapping of past inputs and past outputs to the current output. What is the difference with regard to inputs and outputs between FIR and transfer function (also IIR) linear models?
6. Can FIR systems become unstable (i.e., produce infinitely growing output although the input is finite)?
7. Which mathematical characteristics of a transfer function model are decisive for the type of behavior of linear systems (stable, unstable, or oscillating behavior)? What characteristics of a state-space model correspond to them?
8. What error measure should be minimized in the optimal encoding task (counting to the unsupervised learning task category)?

References

1. Åström KJ, Wittenmark B (1989) Adaptive control. In: Addison-Wesley series in electrical and computer engineering. Addison-Wesley, Reading, MA
2. Cortes C, Vapnik V (1995) Support-vector networks. Mach Learn 20(3):273–297. https://doi.org/10.1007/BF00994018
3. Devlin J, Chang MW, Lee K, Toutanova K (2019) BERT: pre-training of deep bidirectional transformers for language understanding. In: Proceedings of the 2019 conference of the North American chapter of the association for computational linguistics: human language technologies, vol 1 (long and short papers). Association for Computational Linguistics, Minneapolis, MN, pp 4171–4186. https://doi.org/10.18653/v1/N19-1423

4. Duan KB, Keerthi SS (2005) Which is the best multiclass SVM method? An empirical study. In: Oza NC, Polikar R, Kittler J, Roli F (eds) Multiple classifier systems. Lecture notes in computer science. Springer, Berlin, Heidelberg, pp 278–285. https://doi.org/10.1007/11494683_28

5. Enders W (2014) Stationary time-series models, 4th edn. Wiley, Hoboken, NJ, pp 48–107. https://www.wiley.com/en-us/Applied+Econometric+Time+Series%2C+4th+Edition-p-9781118808566

6. Hoffman JD, Frankel S (2001) Numerical methods for engineers and scientists, 2nd edn. Marcel Dekker, New York

7. Hörmander L (2003) The analysis of linear partial differential operators I. Classics in mathematics. Springer Berlin Heidelberg, Berlin, Heidelberg. https://doi.org/10.1007/978-3-642-61497-2

8. Hrycej T (1989) Unsupervised learning by backward inhibition. In: Proceedings of the 11th international joint conference on artificial intelligence, IJCAI'89, vol 1. Morgan Kaufmann Publishers Inc., San Francisco, CA, pp 170–175

9. Hrycej T (1992) Modular learning in neural networks: a modularized approach to neural network classification. Wiley, New York, NY

10. Hrycej T (1997) Neurocontrol: towards industrial control methodology. Wiley, New York, Chichester

11. Koussiouris TG, Kafiris GP (1981) Controllability indices, observability indices and the Hankel matrix. Int J Control 33(4):773–775. https://doi.org/10.1080/00207178108922955

12. Lachenbruch PA (1975) Discriminant analysis. Hafner Press, New York

13. MacQueen J (1967) Some methods for classification and analysis of multivariate observations. In: Proceedings of the fifth Berkeley symposium on mathematical statistics and probability, volume 1: statistics. University of California Press, Berkeley, CA, pp 281–297. https://projecteuclid.org/euclid.bsmsp/1200512992

14. Oppenheim AV, Willsky AS, Young IT (1983) Signals and systems. Prentice Hall, Englewood Cliffs, NJ

15. Pearson K (1901) LIII. On lines and planes of closest fit to systems of points in space. Lond Edinb Dublin Philos Mag J Sci 2(11):559–572. https://doi.org/10.1080/14786440109462720

16. Rosenblatt F (1957) The perceptron—a perceiving and recognizing automaton. Cornell Aeronautical Laboratory

17. Shor NZ (1985) The subgradient method. In: No. 3 in Springer series in computational mathematics. Springer-Verlag, Berlin Heidelberg, pp 22–47. https://doi.org/10.1007/978-3-642-82118-9

18. Smith GD (1985) Numerical solution of partial differential equations: finite difference methods, 3rd edn. In: Oxford applied mathematics and computing science series. Clarendon Press, Oxford University Press, Oxford [Oxfordshire], New York

19. Stützle EA, Hrycej T (2005) Numerical method for estimating multivariate conditional distributions. Comput Stat 20(1):151–176. https://doi.org/10.1007/BF02736128

20. Van Overschee P, De Moor B (1996) Subspace identification for linear systems. Springer US, Boston, MA. https://doi.org/10.1007/978-1-4613-0465-4

21. Vapnik V (2000) The nature of statistical learning theory, 2nd edn. In: Information science and statistics. Springer-Verlag, New York. https://doi.org/10.1007/978-1-4757-3264-1

22. Weibull W (1951) A statistical distribution function of wide applicability. J Appl Mech 18:293–297. https://doi.org/10.1115/1.4010337

Data Processing by Neural Networks

3

In the contemporary DS, artificial neural networks have gained enormous popularity. This is the reason for dedicating a chapter to describing their structure and properties.

The origin of artificial networks has been in mimicking some aspects of neural networks found in living organisms. The motivation for this has been the hope to produce some extent of intelligent behavior. This approach produced interesting achievements in domains such as Natural Language Processing, Computer Vision, and nonlinear forecasting in the last decades. During this time, the discipline of artificial neural networks experienced a development separated from the original natural analogy. The most striking differences are:

- The overwhelming majority of artificial neural networks are implemented as digital devices, while natural neural networks are analog (although their operation may produce a result resembling digital representation such as speech or logical decisions).
- Artificial neural networks exhibit structures fixed before their use, while natural neural networks are typically modified during the organism's life.

It is particularly the latter characteristic that makes artificial neural networks better tractable by analytical methods.

Although there is an enormous variety of applications of artificial neural network technology, most of them can be reduced to a few abstract computational task types. Each of these task types uses a characteristic network structure. In the following sections, these task types and corresponding network characteristics will be discussed. Since this work focuses on artificial (and not natural) neural networks, the attribute *artificial* will be omitted wherever no misunderstanding is possible.

© The Author(s), under exclusive license to Springer Nature Switzerland AG 2023
T. Hrycej et al., *Mathematical Foundations of Data Science*, Texts in Computer Science,
https://doi.org/10.1007/978-3-031-19074-2_3

3.1   Feedforward and Feedback Networks

A neural network is a structure that can be described as a graph consisting of a set
of vertices V and a set of edges E connecting them, as in Fig. 3.1. In the neural
network community, the vertices are frequently referred to as *units* or even *neurons*.
A synonymous term is also *nodes*. Edges are frequently called *connections*.

Although network representations with non-directed graphs can be found in the
literature, we will always use the directed representation in this work, for reasons
explained below. Every edge in Fig. 3.1 has a direction: there is one vertex where
the edge is coming out and one vertex the edge is entering.

To make a neural network an information processing device, some more provisions
are important. During the information processing, the vertices are assigned (most
commonly numerical) values. The way how these values are assigned is the following.
The value of vertex results (in mostly a deterministic way) from the values of all
so-called predecessor vertices connected with it, with an edge directed from the
predecessor toward the vertex concerned. The value of a vertex is a function of the
values of its predecessors. The predominant form of this function is a scalar nonlinear
function of the weighted sum of the predecessor values, modified by an additive
constant called *bias*. In this case, the weights are assigned to individual connections.
It is usual that these weights are fixed by some kind of fitting or *learning* algorithms.
By contrast, the values of vertices are recomputed for every new data input into the
network. In other words, the data processing by the network consists of computing
the values (sometimes called *activations*) of the vertices.

The vertex v_4 of Fig. 3.1 whose only predecessors are v_1 and v_2 would receive
the value

$$v_4 = f\,(w_1v_1 + w_2v_2 + b_4)$$

where f is a usually nonlinear, so-called *activation function*. It would also be possible
to use linear activation functions. This would reduce the network to a linear system
tractable by classical methods of linear algebra and systems theory. The motivation

Fig. 3.1 A set of vertices V
and a set of edges E

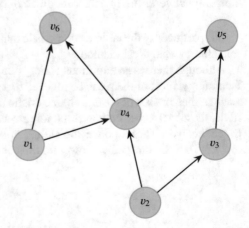

Fig. 3.2 Five popular
activation functions

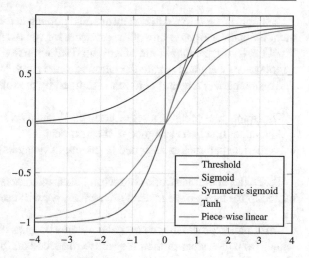

for using nonlinear activation functions is going beyond the limitations of linear
theory.

Let us review some popular activation functions in Fig. 3.2.

Historically, early attempts have been done with a simple threshold function of
the form

$$f(x) = \begin{cases} 1 & \text{if } x > 0 \\ 0 & \text{if } x \leq 0 \end{cases} \tag{3.1}$$

As obvious from the function plot, this function has a discontinuity at $x = 0$
which results in poor tractability by numerical algorithms.

The so-called *sigmoid* function

$$f(x) = \frac{1}{1 + e^{-x}} \tag{3.2}$$

is a kind of continuous analogy to the threshold function. It is frequently used as a
shifted version that is symmetric around the x axis:

$$f(x) = \frac{2}{1 + e^{-x}} - 1 \tag{3.3}$$

or as an additionally scaled version, the tangens hyperbolicus:

$$f(x) = \tanh(x) = \frac{2}{1 + e^{-2x}} - 1 \tag{3.4}$$

A particular property of tangens hyperbolicus is its first derivative being unity at
the point $x = 0$.

Recently, also simpler piecewise linear functions such as *Rectified Linear Units*
(RELU)

$$f(x) = \begin{cases} 0 & \text{if } x \leq 0 \\ x & \text{if } 0 < x < 1 \\ 1 & \text{if } x \geq 1 \end{cases} \tag{3.5}$$

have become widespread.

It should be mentioned that there are also popular activation functions whose argument is not the weighted sum of the predecessor values. An example is the maximum function, taking the maximum of weighted values, or a radial basis function which is a function of weighted square deviations from a *center point* assigned to the vertex.

To summarize, a neural network is defined by the following elements:

- The graph consisting of a set of vertices V and a set of directed edges E;
- Activation functions assigned to the vertices;
- Numerical parameters assigned to the edges (weights) and vertices (bias values).

Since the neural network community has, in contrast to the graph theory, committed to the term *nodes* instead of *vertices*, we will use the former word from now on.

Information processing by neural networks consists of computing node activation values from the values of their respective predecessors, assuming these predecessors already have some values. To do this, some nodes must be assigned initial values before the beginning of the computation. This set of initialized nodes has to be such that for all nodes, there is a path along directed edges from some initialized node (otherwise, some nodes would never receive an activation value). In the example network of Fig. 3.1, the set of initialized nodes has to contain nodes v_1 and v_2—otherwise these nodes would never receive their values.

Under these conditions, the information processing algorithm of such neural networks is genuinely simple. It consists of iteratively performing the following step:

> Compute the activation values of all nodes whose all predecessors already have a value.

For the character of the information processing and the interpretation of its results, it is crucial to differentiate between two cases:

1. Networks containing no closed cycle along the directed edges: *feedforward* networks.
2. Network containing such cycles: *feedback* networks.

It is important to point out that this classification is based on the interpretation of the edge direction as the order in which the activation function is computed: it is computed from the activation values of predecessors (defined by the edge directions).

This has to be considered not to be confused with other interpretations of the edge directions used in some work. A prominent example of a different definition is the so-called *causal* (or *Bayesian*) networks, extensively elaborated by Pearl [5]. In these networks, the edge direction corresponds to the assumed causality, defined by the network designer. The nodes are labeled by some meaningful propositions, such as "rainy day" or "low light intensity outdoors". The proposed causality would probably be that low light intensity will follow from the rainy weather and not inversely.

Nevertheless, causal networks make (probabilistic) inferences in both directions. From the knowledge of a day being rainy, the low light intensity can be forecast. But also, from a measured low intensity, a substantial probability of rainy weather can be concluded. Both directions of conjecture are consistent with common sense and are justified by the Bayesian probability theory and the symmetric relationships between conditional probabilities.

Another remark about the directionality of edges is to be made. For the aims of information processing, a non-directed edge would be substituted by two edges connecting the nodes concerned in both directions. This possibility is the reason why only directed connections are considered here.

3.2 Data Processing by Feedforward Networks

As defined above, *feedforward* networks are those containing no cycles along their directed edges. The network of Fig. 3.1 is an example of such network. To apply the algorithm of Sect. 3.1, the nodes v_1 and v_2, lacking of any predecessors, are to be initialized with activation values. Every other node receives its activation value after a known number of iterations:

- Nodes v_3 and v_4 after one iteration.
- Nodes v_5 and v_6 after two iterations (the evaluation of v_6 being postponed for its predecessor v_4 to have received its value).

After two iterations, the terminal state of the network is attained. Further updates will bring about no changes. The states of all non-initial nodes (vector y) are a nonlinear vector function of the state of initial nodes (vector x):

$$y = f(x)$$

This is an important (and practically advantageous) property of feedforward neural networks:

> For a given input, the information processing of feedforward networks is accomplished in a fixed, finite number of iterations.

The idea of processing within a fixed number of iterations suggests the possibility of a more structured subclass of feedforward networks. This possibility is also supported by some brain structures (e.g., sensomotoric functions in the cerebellum or the image processing in the visual cortex), although their organization is not so strict. The nodes can be organized to layers, defined by the constraint that the nodes of a layer have only predecessors from the preceding layer. Omitting an edge from the network of Fig. 3.1, we receive the network of Fig. 3.3. It corresponds to a layered network in the sense of the mentioned definition, although it is not graphically organized into layers. This makes clear that the property of being "layered" results

Fig. 3.3 A layered network

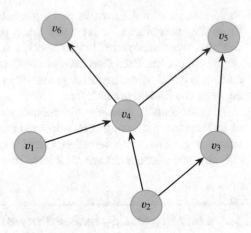

from the predecessor specifications rather than from the graphical representation. Nevertheless, for clarity, layered networks are represented by a sequence of node vectors, each corresponding to a layer.

The (zeroth) input layer consists of nodes v_1 and v_2, the first layer of nodes v_3 and v_4 and the second layer of nodes v_5 and v_6. Input layer, whose nodes are no activations but fixed values, is assigned index zero. The last layer is usually given a prominent function—it is the layer where the result of the computation is displayed. In most application frames, it is also the only layer whose values are given some meaningful interpretation. This is why this layer is called *output layer*. The intermediate layers are usually referred to as *hidden layers*. This organization makes it possible to omit considering individual nodes and observing the whole layers instead. Each layer is, in fact, a vector of activation values. The weights between individual layers can be viewed as a matrix W, the biases are a vector b. For missing edges between nodes of two consecutive layers, the weight matrix entries are set to zero. An example of graphical representation of structures of this type (with two hidden layers) is given in Fig. 3.4.

The ith layer processing is a vector mapping from the activation vector $x_i - 1$ (index of x is i-1) of the $(i - 1)$st layer to the activation vector x_i of the ith layer. In the case of the activation functions of the form

$$F(Wz + b)$$

$f(u)$ is the column vector

$$F(u) = \left[f(u_1) \cdots f(u_m) \right]'$$

(Column vectors are used throughout this work to be consistent with the typical notation of statistics and systems theory.)

Fig. 3.4 Layered network in
vector representation

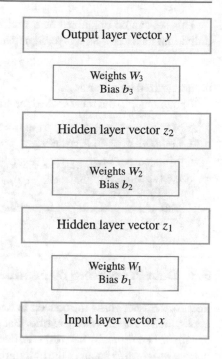

In other words:

1. The column vector of the sums of weighted values of $u = Wx + b$ is computed.
2. The elements of vector u are mapped one by one by the scalar function $g(u_j)$ to the activation vector x_i.

This scheme can be applied to activation functions such as the *sigmoid*, *threshold*, or *piecewise linear* functions using the weighted sum. For other types such as *maximum* function, is has to be modified in an analogous way. In summary, the activation of the ith layer (out of n layers) is

$$z_i = F_i\left(W_i z_{i-1} + b_i\right) \tag{3.6}$$

with $z_0 = x$ (input vector) and $z_n = y$ (output vector).

The processing of the complete feedforward network of Fig. 3.4 representing a mapping G can be formulated as a series of nested functions:

$$y = G(x) = F_n\left(W_n F_{n-1}\left(W_{n-1} F_{n-2}\left(\cdots F_1\left(W_1 x + b_1\right)\right) + b_{n-1}\right) + b_n\right) \tag{3.7}$$

The parameters W_i (weights) and b_i (biases) are kept fixed during the processing. However, they have to be fitted to data for the network to perform in an optimal way. These parameters consist of the set

$$\left\{W_1, b_1, W_2, b_2, \ldots, W_n, b_n\right\} \tag{3.8}$$

This set can be organized as a vector p (e.g., by concatenating the columns of individual matrices). So a layered network can be written as a vector mapping

$$y = G(x, p) \tag{3.9}$$

parameterized by a vector p.

To summarize, the most important statement of this section is the following:

A feedforward network represents a vector function mapping the input vector x to the output vector y. The evaluation is strictly sequential in n time slices corresponding to processing of n layers. It is ultimately accomplished in a finite time.

3.3   Data Processing by Feedback Networks

Feedforward networks introduced in Sect. 3.2 are covering the majority of popular applications. The class of networks that do not belong to the feedforward category is usually called *feedback* networks, although various other names such as *recurrent* or *recursive networks* occur in the literature. They are useful for representing inherently dynamical processes such as spoken language or physical and economical systems. However, their use is substantially more difficult than that of feedforward networks. This is why the user should be informed about the pitfalls of using such networks.

An example of a feedback network can be derived from the feedforward network of Fig. 3.1 by inserting an edge between v_3 and v_4 and inverting the edge between v_3 and v_5. The resulting network is presented in Fig. 3.5. There is an obvious cycle of nodes v_3, v_4, and v_5. Thus, the no-cycle condition for feedforward networks is violated.

Fig. 3.5 A feedback network

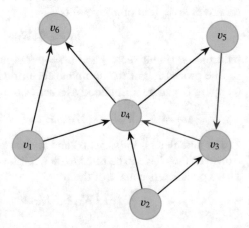

The nodes v_1 and v_2 have no predecessors and have to be assigned initial values. The activation equations of the remaining nodes with corresponding nonzero weights and biases, and activation function f are:

$$
\begin{aligned}
v_3 &= f\left(w_{3,2}v_2 + w_{3,5}v_5 + b_3\right) \\
v_4 &= f\left(w_{4,1}v_1 + w_{4,2}v_2 + w_{4,3}v_3 + b_4\right) \\
v_5 &= f\left(w_{5,4}v_4 + b_5\right) \\
v_6 &= f\left(w_{6,1}v_1 + w_{6,4}v_4 + b_6\right)
\end{aligned}
\tag{3.10}
$$

This is a nonlinear feedback system. Some aspects of such systems have been discussed in Sect. 2.3. The behavior of nonlinear systems is not easy to analyze. The systems theory, which is the mathematical field concerned with system behavior, provides only some (mostly mathematically challenging) statements about the stability of general nonlinear systems. By contrast, there is extensive knowledge concerning linear systems: Their behavior can be described in detail in terms of oscillatory frequencies, reactions to external inputs with known frequency spectrum and other. The analytical and numerical methods for this have reached a substantial degree of maturity. Therefore, it is worth establishing a relationship between nonlinear and linear systems. Setting up an approximation of a nonlinear system by a related linear one opens the door to using powerful analytical instruments available to linear systems, as briefly illustrated below.

The approximation has the following justification. Every nonlinear mapping $g(x)$ can be linearized around a certain point. Taylor expansion provides the tool for approximation of mapping $y = g(x_0)$ in the proximity of the fixed vector x_0. The first two terms of this expansion represent a linear approximation:

$$
y \approx \frac{\partial g(x_0)}{\partial x}(x - x_0) + g(x_0) = Ax + b
\tag{3.11}
$$

The matrix A is the matrix of partial derivatives $a_{i,j} = \dfrac{\partial y_i}{\partial x_j}$ of the vector mapping $y = g(x)$.

For our example network system described in (3.10), the matrix A consists of the weights multiplied by the actual derivative of the activation function $f(u)$ at the point given by its argument, the weighted sum of the predecessor activation values.

The symmetric sigmoid is very close to the linear function $y = \dfrac{u}{2}$ on the interval $[-1, 1]$ and reasonably approximated on the interval $[-2, 2]$ (see Fig. 3.6). This justifies the approximation of (3.10) by

$$
y_{k+1} = \frac{1}{2}W_1 y_k + \frac{1}{2}W_2 u
\tag{3.12}
$$

with u being the fixed input nodes v_1 and v_2 (whose values do not change) and y_k the activation of the remaining nodes (indexed by 3, 4, 5, and 6) at iteration k. For simplicity, bias b is omitted.

Fig. 3.6 Approximation of symmetric sigmoid by a linear function

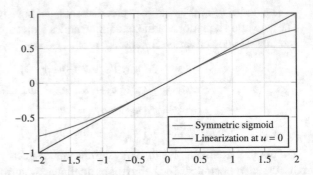

(3.12) can be made more concrete by inserting the edge weights of Fig. 3.5:

$$
\begin{bmatrix} v_3 \\ v_4 \\ v_5 \\ v_6 \end{bmatrix}_{k+1} = \frac{1}{2} \begin{bmatrix} 0 & 0 & w_{3,5} & 0 \\ w_{4,3} & 0 & 0 & 0 \\ 0 & w_{5,4} & 0 & 0 \\ 0 & w_{6,4} & 0 & 0 \end{bmatrix} \begin{bmatrix} v_3 \\ v_4 \\ v_5 \\ v_6 \end{bmatrix}_k + \frac{1}{2} \begin{bmatrix} 0 & w_{3,2} \\ w_{4,1} & w_{4,2} \\ 0 & 0 \\ w_{6,1} & 0 \end{bmatrix} \begin{bmatrix} v_1 \\ v_2 \end{bmatrix} \qquad (3.13)
$$

The activation v_6 obviously depends only on the input node v_1 and the state node v_4, having no successor between other state nodes. So, it is not involved in the feedback loop—it is a pure output node. The state equation corresponding to the difference equation (2.73) with input vector u being $\begin{bmatrix} v_1 & v_2 \end{bmatrix}'$ and state vector $x = \begin{bmatrix} v_3 & v_4 & v_5 \end{bmatrix}'$ is here

$$
x_{k+1} = \begin{bmatrix} v_3 \\ v_4 \\ v_5 \end{bmatrix}_{k+1} = \frac{1}{2} \begin{bmatrix} 0 & 0 & w_{35} \\ w_{43} & 0 & 0 \\ 0 & w_{54} & 0 \end{bmatrix} \begin{bmatrix} v_3 \\ v_4 \\ v_5 \end{bmatrix}_k + \frac{1}{2} \begin{bmatrix} 0 & w_{32} \\ w_{41} & w_{42} \\ 0 & 0 \end{bmatrix} \begin{bmatrix} v_1 \\ v_2 \end{bmatrix} = Ax_k + Bu_k
$$

$$(3.14)$$

The output equation (2.74) with output vector $y = \begin{bmatrix} v_3 & v_4 & v_5 & v_6 \end{bmatrix}$ of all observable nodes is

$$
y_k = \begin{bmatrix} v_3 \\ v_4 \\ v_5 \\ v_6 \end{bmatrix}_k = \frac{1}{2} \begin{bmatrix} 1 & 0 & 0 \\ 0 & 1 & 0 \\ 0 & 0 & 1 \\ 0 & w_{64} & 0 \end{bmatrix} \begin{bmatrix} v_3 \\ v_4 \\ v_5 \end{bmatrix}_k + \frac{1}{2} \begin{bmatrix} 0 & 0 \\ 0 & 0 \\ 0 & 0 \\ w_{61} & 0 \end{bmatrix} \begin{bmatrix} v_1 \\ v_2 \end{bmatrix} = Cx_k + Du_k
$$

This formulation is consistent with the general state-space description of linear dynamical systems, as presented in Sect. 2.3, in corresponding notation.

Let us observe the behavior of the network of Fig. 3.5 described by (3.10) with a symmetric sigmoid activation function and linearized to the form (3.13). The dynamic behavior is determined by the three weights on edges within the cycle $v_3 \rightarrow v_4 \rightarrow v_5$. One of the possible configurations exhibits stable behavior shown in Fig. 3.7. It is a stable behavior with an obvious convergence to fixed values for individual nodes.

Fig. 3.7 Stable behavior

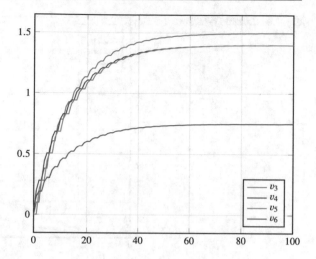

These terminal values (i.e., the steady state of the network) result by solving (3.12) assuming a constant (i.e., no longer changing) activation of state nodes $x_{k+1} = x_k = x$:

$$x = Ax + Bu$$
$$(I - A)\, y = Bu \tag{3.15}$$
$$y = (I - A)^{-1} Bu$$

The stability of the trajectory depends on the eigenvalues of matrix A, as mentioned in Sect. 2.3. Eigenvalues are solutions of equation (2.75), which is equivalent with setting the following determinant to zero:

$$|A - \lambda_i I| = 0 \tag{3.16}$$

Complex eigenvalues are indicators of oscillatory behavior. For the weight configuration, $w_{35} = 2.0$, $w_{43} = 1.6$, and $w_{54} = 2.0$, there are only small oscillations with a period of three iterations.

With a relatively small change of involved weights to $w_{35} = 2.2$, $w_{43} = 1.8$, and $w_{54} = 2.2$, the behavior becomes unstable (see Fig. 3.8). The activations grow to infinity.

Another weight configuration $w_{35} = 1.8$, $w_{43} = 1.8$, and $w_{54} = -1.8$ shows strong oscillations with a mixture of two frequencies with periods of two and six iterations (Fig. 3.9).

Using the sigmoid activation functions makes the system nonlinear. This leads to modifications in behavior. Sigmoid is a bounded function. This prevents its output to grow without limits. In this way, the convergence to some finite values and thus formal stability is enforced. The unstable linear configuration of Fig. 3.8 becomes that of Fig. 3.10. However, this *stability* guarantee is mostly not helpful for the analysis. It is not simply a *stabilized version* of the linear counterpart—even the signs of the activations do not correspond between the linear and the nonlinear cases. Although saturation effects lead to a less spectacular adverse state of the application, it is

Fig. 3.8 Unstable behavior

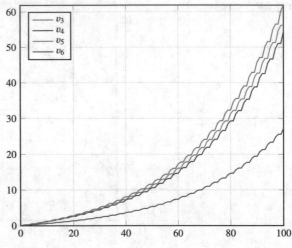

Fig. 3.9 Oscillating behavior

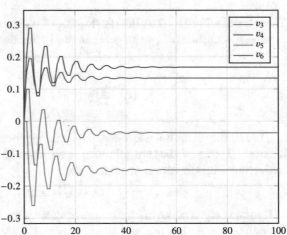

mostly characterized by nodes being trapped in the saturation state from which there is no way back.

The complexity of the behavior of nonlinear systems is not limited to the properties stated. If the input is variable, for example, in the form of a sequence of vector patterns, the response of the network nodes is delayed. The delay depends on the variability of pattern sequences. In the terminology of dynamic systems, it is the phase of the signal transfer. The phase may vary from node to node, depending on its connectivity to other nodes. This may play a significant role in applications with a continuous flow of input data such as speech processing.

Fig. 3.10 Behavior of a network whose linear approximation is unstable

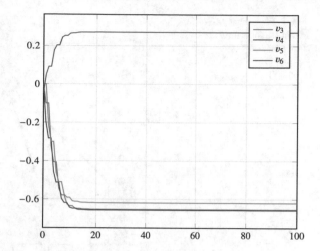

To summarize the statements of this section:

> Feedback networks exhibit complex behavior in time. Their reaction to a static input persists over infinite time. The most important adverse behavior characteristics are potential instability which may lead to a complete failure of the application. The oscillatory behavior is also mostly not desirable, except for cases in which it is the focus of the application. The analysis of the behavior can be supported by linear analogies. .

3.4 Feedforward Networks with External Feedback

In addition to some difficult facets of the behavior of feedback networks inherent to any feedback, there are some burdens for the user that may make their application challenging.

The first of them is that it is not easy to select their structure (in particular, the connections) following justified principles.

Another one concerns the data for determining the parameters. A usual concept for feedforward networks is parametrization with the help of pairs of input and output vectors. The former are simply to be mapped to the latter.

There is no such simple principle for feedback networks. The reason for this is their infinite response even to a static signal. It is not practical to provide infinite or nearly infinite output sequences for every input vector.

Fortunately, feedback systems can always be expressed in a form inspired by the transfer function concept discussed in Sect. 2.3. A neural network is then a direct representation of a transfer function, with additional possibility of capturing nonlin-

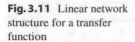

Fig. 3.11 Linear network structure for a transfer function

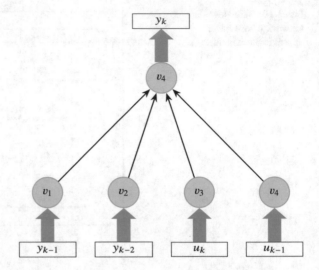

earities. To understand the statements of the present section, the comprehension of Sect. 2.3 is required.

The linear transfer function scheme corresponds to the network structure of Fig. 3.11. The coefficients a_i and b_i from (2.76) correspond to the weights assigned to the corresponding edges. Obviously, this is a feedforward structure that can be parameterized by input/output pairs. However, there is a special feature of these input/output pairs: they contain the same values at different positions. For example, the value y_{100} will occur as an output value in one input/output pair $(y_{99}, y_{98}, u_{100}, u_{99} \Rightarrow y_{100})$ and in two additional consecutive input/output pairs $(y_{100}, y_{99}, u_{101}, u_{100} \Rightarrow y_{101})$ and $(y_{101}, y_{100}, u_{102}, u_{101} \Rightarrow y_{102})$ as delayed input values.

The system represented by this structure contains outer feedback through delayed values. Its operation follows the structure of Fig. 3.12. Black arrows are usual connections without delay, as in Fig. 3.11. Additional red arrows correspond to feedback with variable delay by one period. The only genuine external input is u_k.

This structure is easy to parameterize by methods usual for feedforward networks. The training set is received by collecting the input/output pairs with various delays. However, the behavioral complexity inherent to feedback systems remains. The stability and tendency to oscillate result in the same way from some characteristics, presented in Sect. 2.3. As stated there, the eigenvalues of the state-space system representation mentioned in Sect. 3.3 correspond to the system poles.

To remind of the concept of poles defined by (2.79), they are the roots of the polynomial in delay operator z with coefficient a_k being the weights of delayed output variables:

$$z^m - \sum_{i=1}^{m} a_k z^{m-i} \tag{3.17}$$

(The weights of delayed or non-delayed input variable are irrelevant for the poles.)

Fig. 3.12 Outer feedback to a linear feedforward structure

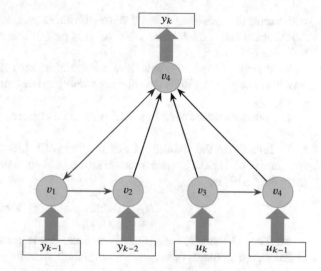

As stated in Sect. 2.3, the role of poles is equivalent to that of eigenvalues in determining the stability and oscillatory tendency. Poles with absolute values less or equal to unity guarantee stability, the occurrence of larger absolute values lead inevitably to unstable behavior. The oscillatory response is possible with complex poles.

Besides determining poles from (3.17), representing the network in the state-space form (2.73) and (2.74) is possible. For the configuration of Fig. 3.11, the recursive relationship is:

$$\begin{bmatrix} y_{d0} \\ y_{d1} \\ u_{d1} \end{bmatrix}_k = \begin{bmatrix} a_1 & a_2 & b_1 \\ 1 & 0 & 0 \\ 0 & 0 & 0 \end{bmatrix} \begin{bmatrix} y_{d0} \\ y_{d1} \\ u_{d1} \end{bmatrix}_{k-1} + \begin{bmatrix} b_0 \\ 0 \\ 1 \end{bmatrix} u_{d0} \tag{3.18}$$

The indices of the form di of output y and input u symbolize the delay—the variable y_{d1} is the output delayed by one period. The index $_{d0}$ denotes the non-delayed values of input and output. The first row of matrices A and B are filled by the weights representing the input/output mapping (without outer feedback). The following rows of A and B represent the feedback delays. For example, the new value (kth period) of y_{d1} is identical to the old value (($k - 1$)th period) of y_{d0}.

The eigenvalues of matrix A are identical with the poles of the polynomial (3.17). The advantage of this representation is that it can be generalized to the case of multiple inputs and multiple outputs. The stability and oscillation analysis is identical to that of Sect. 3.3.

So far, the linear case has been considered because of its transparency and analytical tractability. However, a substitution of the linear network from Fig. 3.11 by a layered structure of Fig. 3.4 is straightforward. This structure represents the non-linear mapping of the type $y = G(x)$ from (3.7), parameterized as in (3.9). In this case, the mapping arguments are modified to

$$y_k = G([v_k, u_k]) \tag{3.19}$$

with vector v_k containing delayed values of output y_{k-d} and input u_{k-d} with $d > 0$. Additionally, the non-delayed input u_k is a part of the concatenated argument of mapping G.

Most real feedback systems have a fixpoint at zero (for zero input, the output should also be zero). This is why bias constants can frequently be omitted in feedback systems.

The outer feedback as in Fig. 3.12 is handled correspondingly for nonlinear systems.

To investigate the stability of this mapping if subject to outer feedback of the structure from Fig. 3.12, a linearization around a given state can be used. A linearized mapping (3.19) is:

$$y_k = \frac{\partial G}{\partial v_k} v_k + \frac{\partial G}{\partial u_k} u_k = G_v v_k + G_u u_k$$

Matrices G_v and G_u are Jacobian matrices, that is, matrices of partial derivatives of individual elements of vector mapping G about individual elements of argument vectors v and u.

The feedback state-space matrices A and B are constructed analogous to (3.18):

$$A = \begin{bmatrix} G_v \\ H_v \end{bmatrix} \quad B = \begin{bmatrix} G_u \\ H_u \end{bmatrix}$$

with submatrices H_v and H_u describing the delay relationships.

The eigenvalues of A determine the stability at the given network state described by the activations of all its nodes. For a strict proof of stability, this would have to be done for all (infinitely many) activation states, which is infeasible. In the best case, a small selection of important states can be investigated. With sigmoid activation functions (and other activation functions with the largest gradient at zero), the state with the largest instability potential is the state of all nodes being zero. The reason for this is that the derivative of a sigmoid is the largest with the sigmoid argument being zero. So, the stability investigation should take place at least at this working point.

It is important to point out that linear instability around one state results in *overall instability of the nonlinear system*. So wherever unstable eigenvalues are found, the network is to be classified as unstable, which, in turn, makes the network useless for applications.

For large networks with thousands of input and output variables, the Jacobian matrices of partial derivatives are of considerable size ($m \times n$ matrix elements for m inputs and n outputs). However, the computation is easy with the help of backpropagation formulas for gradient computation. The next step is to complete the system matrix with delay relationships (submatrices H above). Then, the eigenvalues are to be computed. Most numerical packages (e.g., *SciPy*[1]) provide procedures for efficient computation of eigenvalues even for matrices with hundreds of thousands to millions of elements.

---

[1] https://www.scipy.org.

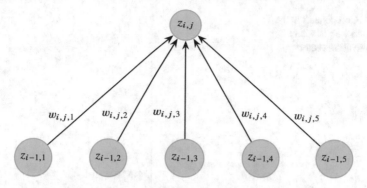

Fig. 3.13 Weights on the connections of a node with the preceding layer

To summarize:

Feedforward layered networks with outer feedback are transparent alternatives for feedback networks. They can be parameterized in a way usual for feedforward networks: with the help of input/output data pairs.

However, they exhibit complex dynamic behavior analogous to feedback network (an all feedback systems), with a potential for instability. Instability potential can be investigated with the help of linearization, to prevent parametrizations useless for most applications.

A possible remedy for stability problems is the use of a multistep error measure mentioned in Sect. 2.3.

3.5 Interpretation of Network Weights

In most cases, network weights are received by an algorithm for fitting the network to the data, as discussed in Chap. 5. Many successful recent applications use networks with hundreds of thousands to millions of weights. In such cases, meaningful interpretation of network weights cannot be expected.

Nevertheless, the basic principles behind the weights are important to understand. They are helpful for the design of appropriate network structures. An example are the convolutional networks introduced in Sect. 3.6.

Let us consider the connections of a node of the ith layer with all nodes of the preceding layer (indexed by $i - 1$), as shown in Fig. 3.13.

The activation of the jth node of the ith layer $z_{i,j}$ is

$$z_{i,j} = f\left(w_{ij}z_{i-1}\right) \tag{3.20}$$

The argument of this activation function f is the scalar number $w_{ij}z_{i-1}$. This is a product of the jth row w_{ij} of the weight matrix W_i of ith layer and the activation

Fig. 3.14 Components in the
product of a weight vector
and an activation vector

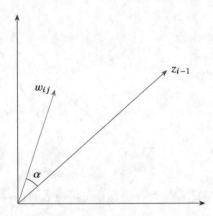

column vector z_{i-1} of the $(i - 1)$st layer. Both factors are vectors, so the term
corresponds to a vector product. There is a well-known relationship between a vector
product xy and the angle between both vectors:

$$xy = \cos(\alpha) |x| |y| \tag{3.21}$$

Absolute values of vectors x and y correspond to the lengths of the vectors in
Euclidean geometric space:

$$|x| = \sqrt{\sum_i^I x_i^2}$$

This relationship can be applied to the product of the weight vector and the activation vector from (3.20):

$$w_{ij} z_{i-1} = \cos(\alpha) |w_{ij}| |z_{i-1}| \tag{3.22}$$

For two-dimensional vectors (an unrealistic but illustrative dimensionality), the
relation can be depicted as in Fig. 3.14.

According to (3.22), the argument of the activation function f is proportional to

1. the length of the weight vector;
2. the length of the activation vector of the preceding layer;
3. and the cosine of the angle between both vectors.

After the parametrization phase of the network, the weight vector is fixed while
the activation vector is variable, depending on the data processed by the network.
So, two factors affect the magnitude of the argument of the activation function:

1. the cosine which is maximum (equal to unity) if the activation vector points
   toward the same direction as the weight vector and
2. the length of the activation vector.

To summarize, the activation of the ith layer node grows with (1) the similarity of the preceding layer activation vector with the weight vector and (2) the magnitude of the activation vector.

This can be interpreted as a simple way of *recognizing a pattern* in the preceding activation layer. The activation of a node show the extent to which the pattern described by the weight vector can be found in the preceding layer activation vector. In intermediary, hidden layers, this amounts to the extraction of pattern features that are hopefully important for the succeeding layers and finally the output layers.

The whole information processing of a network consists of successively extracting features from the data. The feature patterns to be extracted are defined by the rows of the respective weight matrix.

3.6   Connectivity of Layered Networks

In a layered network, each layer represents a mapping as in (3.6). The network as a whole is then a nested mapping (3.7), in which the output of each layer becomes the input of the next layer, according to the scheme in Fig. 3.4.

The application of weights between the layers can be represented by a matrix W_i. This matrix may or may not contain zero entries. The case in which such zero entries are chosen deliberately corresponds to a network whose layers are not completely connected—zero weights are, in fact, missing connections. This property is frequently referred to as *connectivity*.

There are many possible selections of incomplete connectivity. Most of them have no particular sense, but there is an outstanding case with an excellent success record in some application domains, the so-called *convolutional layers*.

The idea of this connectivity structure is that each unit of the ith layer, that is, each element z_{ij} of vector z_i in the notation of Fig. 3.4, is connected only with the elements of vector z_{i-1} belonging to a certain fixed "environment" of z_{ij}.

The concept of an environment presumes the following properties:

• The layer vector has to possess topological interpretation in which it makes sense to speak about topological neighborhood. This is the case, for example, along a timeline, where consecutive time slices are neighbors of each other. Another example is a pixel image in which the neighborhood is defined by the vicinity of pixels.
• This topological interpretation has to be common for both layers so that the vicinity of z_{ij} can be determined in layer z_{i-1}.

A trivial example of such a layer is given in Fig. 3.15. Every node z_{ij} of the ith layer is connected to the corresponding node $z_{i-1,j}$ of the $(i-1)$st layer as well as to

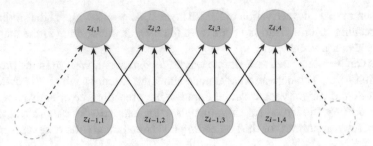

Fig. 3.15 Connectivity constrained to a topological environment

its two immediate neighbors $z_{i-1,j-1}$ and $z_{i-1,j+1}$. The connections at both margins
are truncated to keep the indexing of both layers consistent, but the dotted nodes may
be included to make the marginal connections analogous to the inner nodes. In that
case, the $(i-1)$st layer would have two additional nodes.

As explained in Sect. 3.5, the weight vector assigned to the node expresses the
feature to be extracted from the activation vector of the preceding layer. In a configuration of the type depicted in Fig. 3.15, the feature pattern is confined to connected
nodes. For layers reflecting some real-world topology, an important possibility is
suggested: the feature may be invariant to the node position. In other words, every node of the ith layer extracts the same feature type from its environment in the
$(i-1)$st layer. To do this, all weight vectors (rows of the weight matrix) within this
layer are to be identical except for the index shift. For the configuration of Fig. 3.15,
the weight matrix would be

$$W_i = \begin{bmatrix} w_0 & w_1 & 0 & 0 \\ w_{-1} & w_0 & w_1 & 0 \\ 0 & w_{-1} & w_0 & w_1 \\ 0 & 0 & w_{-1} & w_0 \end{bmatrix} \tag{3.23}$$

using only three distinct weights w_{-1} (for preceding layer node with an index lower
by 1, i.e., one position to the left), w_0 (for preceding layer node with the same index,
i.e., the node "below" the actual layer unit) and w_1 (for preceding layer node with
an index higher by 1, i.e., one position to the right). Note that this is a substantial
reduction compared to fully connected layers with 16 weights. This ratio becomes
even more substantial for realistically long layer vectors—even for a layer node
vector of 1000 units, the number of distinct weights would remain to be three.

The fact that the activation of the jth node of the ith layer is computed as

$$z_{i,j} = f\left(\begin{bmatrix} w_{-1} & w_0 & w_1 \end{bmatrix} \begin{bmatrix} z_{i-1,j-1} & z_{i-1,j} & z_{i-1,j+1} \end{bmatrix}' \right)$$

$$= f\left(w_{-1}z_{i-1,j-1} + w_0 z_{i-1,j} + w_1 z_{i-1,j+1} \right)$$

or, generally, with the symmetric environment of width $2K+1$

$$z_{i,j} = f\left(\sum_{k=-K}^{K} w_k z_{i-k,j-k} \right)$$

Fig. 3.16 First-derivative operator—a change detector

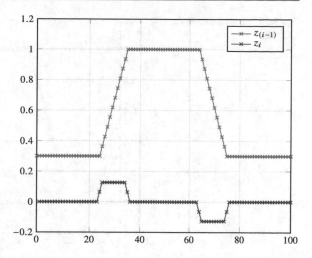

motivates the term "convolutional layer"—the summed term has the form of a mathematical convolutional operator.

A weight structure organized in this way represents a *local operator* (local with regard to the given topology). As explained in Sect. 3.5, every weight vector can be viewed as a "pattern matcher", detecting a certain pattern. The activation of the node with these weights is growing with the growing similarity between the preceding layer activation vector and the weight vector. In the case of a local operator, this activation is growing in the same way for all nodes of the layer: it detects the same pattern at every position.

An example of a simple detector is detecting the gradient in the given environment. A symmetric weight vector for this is

$$[w_{-1} \; w_0 \; w_1] = [-1 \; 0 \; 1] \tag{3.24}$$

The vector product with the relevant segment of the node activation vector

$$[-1 \; 0 \; 1][z_{i-1,j-1} \; z_{i-1,j} \; z_{i-1,j+1}]' = z_{i-1,j+1} - z_{i-1,j-1} \tag{3.25}$$

which corresponds to the first difference, the discrete analogy to the first derivative. This term is positive for values growing with j, negative for decreasing values, and zero for constant ones.

If the vector z_{i-1} corresponds, for example, to a time sequence of values, changing values are detected in the next layer z_i. This situation is depicted in Fig. 3.16 with individual activations corresponding to crosses. It is obvious that wherever the activation in layer $i-1$ changes along the layer vector index, the activation in layer i grows. Constant segments along layer $i-1$ lead to zero activation in layer i.

To detect a "bend", the local weight vector

$$[w_{-1} \; w_0 \; w_1] = [1 \; -2 \; 1] \tag{3.26}$$

with the vector product

$$[1 \; -2 \; 1][z_{i-1,j-1} \; z_{i-1,j} \; z_{i-1,j+1}]' = z_{i-1,j+1} - 2z_{i-1,j} + z_{i-1,j-1} \tag{3.27}$$

Fig. 3.17 Second-derivative operator

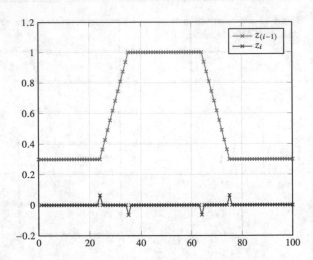

implements the second difference, the discrete counterpart of the second derivative. A simple example of such processing is given in Fig. 3.17.

This type of convolutional operator in a single dimension is useful for data vectors with a uni-dimensional environment. Typical for this is a time axis. This suggests applications with temporal signals such as speech recognition.

Other applications may have multidimensional environments. A prominent example is the 2D image processing.

In an image topology, the environment of a pixel is defined in two dimensions since the neighbors of a pixel extend along two axes. However, the operators on this topology follow analogical principles and have effects similar to those of one-dimensional topologies. An analogy to the change rate operator of (3.26) and (3.27) is the convolutional matrix operator

$$\begin{bmatrix} w_{-1,-1} & w_{-1,0} & w_{-1,1} \\ w_{0,-1} & w_{0,0} & w_{0,1} \\ w_{1,-1} & w_{1,0} & w_{-1,1} \end{bmatrix} = \begin{bmatrix} 1 & 0 & 1 \\ 0 & -4 & 0 \\ 1 & 0 & 1 \end{bmatrix} \tag{3.28}$$

It is applied to the environment of pixel p_{jk} (here without layer index i):

$$\begin{bmatrix} p_{j-1,k-1} & p_{j-1,k} & p_{j-1,k+1} \\ p_{j,k-1} & p_{j,k} & p_{j,k+1} \\ p_{j+1,k-1} & p_{j+1,k} & p_{j+1,k+1} \end{bmatrix} \tag{3.29}$$

The operator (3.28) is applied to the environment (3.29) in a way analogical to the vector product: multiplying the corresponding elements one by one and summing the products. In the case of operator (3.28), the result is:

$$p_{j-1,k-1} + p_{j-1,k+1} + p_{j+1,k-1} + p_{j+1,k+1} - 4p_{j,k} \tag{3.30}$$

The operator (3.28) is a discrete analogy of the derivative along two dimensions, the widespread Laplace operator

$$\nabla^2 f = \frac{\partial^2 f}{\partial x^2} + \frac{\partial^2 f}{\partial y^2} \tag{3.31}$$

Fig. 3.18 Two-dimensional array

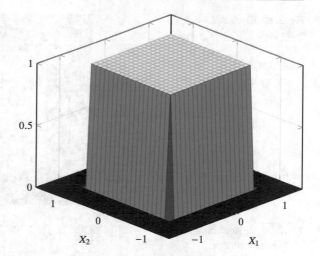

Fig. 3.19 Laplace operator array

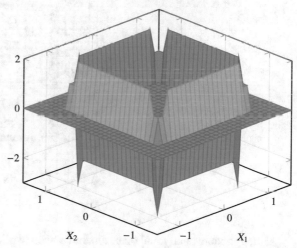

Suppose the pixel values (possibly gray intensity values) in an image are as in the 3D plot in Fig. 3.18.

The array of values computed according to (3.30) is shown in Fig. 3.19.

Wherever there are points with fast intensity change, the Laplace operator receives high positive or negative values. This can be interpreted as "edge detection", an important operation in image processing. Another view of the pixel array of Fig. 3.18, with different pixel values symbolized by black and white color is presented in Fig. 3.20 and the absolute values of Laplace operators in Fig. 3.21.

Figures 3.20 and 3.21 illustrate the edge detection capability of the operator.

Fig. 3.20 Black and white
pixel image

Fig. 3.21 Edge extraction
by *Laplace* operator

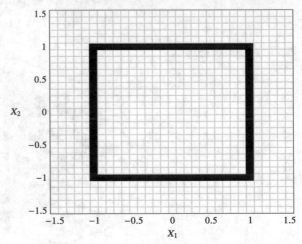

Although many analytical operators have been explicitly proposed, it is also possible to represent them in a neural network with the potential of optimizing them
for a certain image processing task with the help of data. To transform a 2D pixel
array to a layer activation vector, the matrix of pixel values is to be reorganized by
concatenating the row or column vectors to a layer vector. In the case of column
pixel vectors p_i organized into the matrix $\begin{bmatrix} p_1 & \cdots & p_K \end{bmatrix}$, the transformation to the ith
layer vector is

$$\begin{bmatrix} p_1 & \cdots & p_K \end{bmatrix} \rightarrow z_i = \begin{bmatrix} p_1 \\ \vdots \\ p_K \end{bmatrix} \tag{3.32}$$

We can conclude the following:

> Network layers with selective (frequently sparse) connections and correspond-
> ing weights are appropriate for implementing local operators on layers with
> topological meaning with a defined neighborhood. Besides the restricted con-
> nectivity, their fundamentally beneficial property is the identical parametriza-
> tion for every instance (defined by its center element p_{jk} above) of the neigh-
> borhood. This leads to (1) a consistent realization of local operators and (2)
> parameter economy—very large layers can be parameterized by few weights.
>
> However, applications without such meaningful topological organization
> are no candidates for the beneficial use of such network architectures.

3.7   Shallow Networks Versus Deep Networks

In the early times of artificial neural networks, the networks initially consisted of
only one layer. The famous *Perceptron* has only direct weighted connections between
the input and output. The focus at that time was on finding learning rules for simple
tasks such as separating two classes of patterns.

The surge of interest in the eighties has been motivated by discovering the possibility to train networks with more than a single layer, that is, with at least one hidden
layer in the sense of Sect. 3.2. Although more than one hidden layer occurred, too,
the motivation for using them was not clear. The theoretical basis was the theorem of
Cybenko [3], stating that a neural network with a single hidden layer and sigmoid activation functions can approximate an arbitrary continuous function. Unfortunately,
no limit for the number of hidden nodes has been found—depending on the function
form, this number may be infinite.

The recent revival of intensive investigations of neural networks experienced the
shift of focus to networks with multiple hidden networks. The popular term *Deep
Learning* (DL) refers to this property—the structure with many successive layers can
be viewed as "deep". Consistently with this terminology, we will refer to networks
with a single hidden layer as shallow, to those with more than one hidden layer as
deep.

Deep networks have brought about some excellent results in various application
domains. In some cases, these were large networks with millions of weights.

There seems to be no doubt that specific deep architectures such as those using convolution networks (mimicking spatial operators in image processing) are very useful
for specific problems. Sometimes, a whole sequence of such operators is necessary
to extract powerful features for the given applications. Such operator sequences have
been used in image processing already before the surge of deep networks.

The enthusiasm arisen from the success record of such networks seemed to support the hypothesis that the architectural depth is fundamentally superior to shallow

networks. Besides some experimental findings supporting deep network representational efficiency, there are also several attempts for theoretical justifications.

Their hypothesis is essentially the following: Deep networks exhibit larger representation capacity than shallow networks. That is, they are capable of approximating a broader class of functions with the same number of parameters.

To compare the representation capacity of deep and shallow networks, several interesting results have been published. Bengio et al. [1] have investigated a class of algebraic functions that can be represented by a special structure of deep networks (alternating summation and product layers). They showed that for a certain type of deep networks the number of hidden units necessary for a shallow representation would grow exponentially while in a deep network it grows only polynomially.

Montúfar et al. [4] have used a different approach for evaluating the representational capacity. They investigated how many different linear hyperplane regions of input space can be mapped to an output unit. They derived statements for the maximum number of such hyperplanes in deep networks, showing that this number grows exponentially with the number of hidden layers. The activation units used had special, non-smooth activation functions: so-called *rectified linear units* (ReLU) and *softmax* units.

It is common to these findings that they do not make statements about arbitrary functions. The result of Bengio et al. [1] is valid for algebraic functions representable by the given deep network architecture, but not for arbitrary algebraic terms. Montúfar et al. [4] have derived the maximum number of representable hyperplanes but it is not guaranteed that this maximum can be attained for an arbitrary function to be represented. In other words, there are function classes that can be efficiently represented by a deep network while other functions cannot.

This is not unexpected: knowing that some N_1 dimensional function is to be identified from N_2 training examples (which is equivalent to satisfying $N = N_1 N_2$ equations), it cannot be generally represented by less than N parameters although cases representable by fewer parameters exists.

A familiar analogy is that of algebraic terms. Some of them can be made compact by the distributive law, others cannot.

The term:

$$ae + af + ag + ah + be + bf + bg + bh + ce + cf + cg + ch + de + df + dg + dh$$

can be economically expressed as

$$(a + b + c + d)(e + f + g + h)$$

reducing the originally sixteen summands to eight.

However, this cannot be done to an equally economical extent with the term

$$ae + af + ag + ah + be + bf + bg + bh + ce + cf + cg + ch + de + df + dg + h$$

This insight can be summarized in the following way. There exist mappings that can be represented by deep neural networks more economically than by shallow ones. These mappings are characterized by multiple uses of some intermediary (or hidden) concepts. This seems to be typical for cognitive mappings, so that deep networks may be adequate for cognitive tasks. On the other side, the same statement about general functions has not been proven.

The systematic study of Bermeitinger et al. [2] has not confirmed the superiority of deep (with three to five hidden layers) networks over shallow networks with a single hidden layer. For randomly generated mappings, the approximation quality was systematically better with shallow networks with an almost identical number of weight parameters.

This can be attributed to the fact that the convergence of learning algorithms (see Chap. 5) is substantially better for shallow networks.

> Deep networks with dedicated layers extracting special features important for the applications are frequently superior to fully connected shallow networks. This could not so far be confirmed for fully connected deep networks with a comparable number of weight parameters.

3.8 Chapter Summary

Neural networks are a particular type of parameterized nonlinear mapping. They can be used for all tasks mentioned in Chap. 2. Two classes of neural networks with fundamentally different properties are

- feedforward networks and
- feedback networks.

Feedforward networks are frequently organized into layers: the resulting mapping is then a nested function made of mappings represented by individual layers. This nested function can be unanimously evaluated for any input pattern.

By contrast, feedback networks correspond to dynamical systems. Their response to an input pattern is potentially infinite, i.e., they produce an infinite sequence of output patterns. Their dynamical character implies the possibility of unstable or oscillating behavior. It is a difficult task to prevent instabilities during the network learning.

Nonlinear dynamical systems can also be represented by a feedforward network using external feedback: past network outputs are included in the input pattern. This arrangement is equivalent to, but substantially more transparent than using feedback networks.

Beside feedforward networks with completely connected adjacent layers (having connections between all units of one layer and all units of an adjacent layer), reduced connectivity is meaningful and computationally advantageous for special goals. A prominent example are convolutional networks with connections restricted to the close topological environment of the units (e.g., representing image pixels and their environment).

3.9   Comprehension Check

1. A neural network can be represented by a directed graph made of nodes and connecting edges. What is the condition for the network to be a "feedforward network"?
2. Which basic behavior types can be exhibited by feedback networks?
3. Which mathematical characteristics of a neural network (linearized at a certain state of node activation) determine the stability, instability or oscillatory behavior?
4. How would instability of a (linearized) network manifest itself if the activation functions have saturating limits (e.g., a sigmoid function)?
5. How can the vector of weights entering a unit be interpreted with regard to unit's input vector?

References

1. Bengio Y, Delalleau O, Roux NL, Paiement JF, Vincent P, Ouimet M (2004) Learning eigen-functions links spectral embedding and kernel PCA. Neural Comput 16(10):2197–2219. https://doi.org/10.1162/0899766041732396
2. Bermeitinger B, Hrycej T, Handschuh S (2019) Representational capacity of deep neural networks: a computing study. In: Proceedings of the 11th international joint conference on knowledge discovery, knowledge engineering and knowledge management—volume 1: KDIR. SCITEPRESS—Science and Technology Publications, Vienna, Austria, pp 532–538. https://doi.org/10.5220/0008364305320538
3. Cybenko G (1989) Approximation by superpositions of a sigmoidal function. Math Control Signal Syst 2(4):303–314. https://doi.org/10.1007/BF02551274
4. Montúfar GF, Pascanu R, Cho K, Bengio Y (2014) On the number of linear regions of deep neural networks. In: Ghahramani Z, Welling M, Cortes C, Lawrence ND, Weinberger KQ (eds) Advances in neural information processing systems 27. Curran Associates, Inc., pp 2924–2932. http://papers.nips.cc/paper/5422-on-the-number-of-linear-regions-of-deep-neural-networks.pdf
5. Pearl J (1988) Probabilistic reasoning in intelligent systems. Morgan Kaufmann

Learning and Generalization

4

The mission of DS is to extract interesting information from data. In the previous chapters, the form of this information has been discussed: These are diverse types of mappings such as classifiers or quantitative models. In the present chapter, the requirements on data available for processing and discovering this information are reviewed.

In typical recent applications, these data are of considerable size. They may cover thousands or even millions of measured patterns or observed cases. Such quantities can only be processed by automatic algorithms—viewing individual patterns by experts is not feasible.

For serious applications, it is important to assess the reliability of the data analysis results. This is why the most of them follow the approach of *supervised* learning. This learning concept extracts interesting associations between two patterns: an input pattern and an output pattern. The gained information is represented by a mapping of the former to the latter. In some cases, both patterns can be numeric data vectors. In others, the output is the assignment of the input pattern to a class or to a semantic meaning. Figuring out the mapping, a supervised learning approach requires a set of such pattern pairs. Formally, these are pairs of input and output vectors. Once the mapping is known, it can be applied to an input pattern alone, receiving the corresponding output pattern. In this supervised learning framework, the congruence of the received mapping with the given empirical data can be directly quantified by error measures.

The set of input/output pairs used for finding the model optimally congruent with the data is frequently referred to as a *training set*. The procedure of determining the mapping from the training set is called *learning* in the communities working with neural networks and AI. Statistical approaches use more traditional terms such as *regression*. The meanings of both are overlapping to a large degree.

For real-world applications, a mapping consistent only with the training set is of limited interest. The main goal is the forecast of output given a novel input. The forecast may consist of assigning the class to the input pattern (e.g., recognizing an object in an image or determining the semantics of an ambiguous phrase) or making a real-valued forecast of some future parameters such as economic growth.

© The Author(s), under exclusive license to Springer Nature Switzerland AG 2023
T. Hrycej et al., *Mathematical Foundations of Data Science*, Texts in Computer Science,
https://doi.org/10.1007/978-3-031-19074-2_4

The crucial question is how reliable this forecast is. To assess the credibility of the output pattern forecast, another set of input/output patterns, the *test set*, is frequently used. The mapping determined from the training set is applied to the input patterns of the test set and the forecasts are compared with the corresponding output patterns. Forecasting for patterns not included in the training set is called *generalization*. This comparison gives the user a picture of the forecast reliability.

This chapter focuses on the questions concerning the attainable error for the training set and justified expectations about generalization. The majority of mathematically founded statements concern linear mappings. Nevertheless, nonlinear mappings can frequently be analyzed with the help of linearization and so benefit from the results for linear mappings. This relationship is discussed in detail in Sect. 4.2.

4.1   Algebraic Conditions for Fitting Error Minimization

The process of finding an appropriate mapping of input vectors to output vectors with the help of a training set of K pairs (x_k, y_k) consists in determining parameters for which the mapping makes exact (or the best) output forecasts.

Suppose the set of K input column vectors (each of length N) is organized in the $N \times K$ matrix

$$X = \begin{bmatrix} x_1 \ldots x_K \end{bmatrix} \tag{4.1}$$

and the set of corresponding K output column vectors (each of length M) in the $M \times K$ matrix

$$Y = \begin{bmatrix} y_1 \ldots y_K \end{bmatrix} \tag{4.2}$$

For a linear mapping $y = Bx$ to make an exact forecast of all output vectors of the training set, it has to satisfy the matrix equation

$$Y = BX \tag{4.3}$$

This matrix equation can be viewed as M separate equation systems for M row vectors of Y and B:

$$y_m = b_m X, \quad m = 1, \ldots, M \tag{4.4}$$

Each of these equation systems consists of K linear equations (for each of K elements of y_m) for N unknown variables b_{mn} (elements of the mapping vector b_m).

Thus, all M equation systems of (4.3) constitute a system of MK linear equations for MN unknown variables (elements of matrix B, i.e., unknown parameters of the mapping $y = Bx$).

The conditions under which such linear equation systems can be solved are known from basic linear algebra. An equation system $Az = c$ (or its transposed form $z'A' = c'$) has a single, unambiguous solution for the variable vector z if matrix A is square and of full rank. The former condition is equivalent to the requirement of the number of unknown variables being equal to the number of equations. The latter condition states that neither the rows nor the columns of matrix A are linearly dependent.

Besides the case of square matrix A, assuming full rank, the following alternatives exist:

- If there are more equations than variables, there is no exact solution of this equation system for variable vector z. This is the *overdetermined* case.
- If there are fewer equations than variables, there are infinitely many solutions. This is the *underdetermined* case.

These alternatives also apply to the matrix equation (4.3). With a square input data matrix X of full rank, the solution for the mapping matrix B is

$$B = YX^{-1} \tag{4.5}$$

The matrix inversion is feasible whenever matrix X is square and of full rank.

However, this case is rather scarce in practice. The length N of input patterns is usually determined by the available measurement set. The size K of the training set is limited by available data for which true or measured output patterns can be supplied. If many such training pairs have been recorded, it is reasonable to use all or most of them to make the training set as representative for the scope of existing variants as possible. So, both cases of unequal numbers of MK linear equations and MN unknown variables are frequently encountered.

Fortunately, for both cases, there are algebraic solutions. In the overdetermined case (i.e., $N < K$), a model that exactly hits all output vectors does not exist. Instead, a parametrization which is the best approximation in some appropriate way has to be sought. The criterion for the approximation quality will usually have to do with the difference between the forecast and the measured or supplied training output. For each element of the output vector, this will be

$$e_m = b_m X - y_m, \quad m = 1, \dots, M \tag{4.6}$$

From various possibilities to summarize the deviation vector e_m in a single figure, the sum of squares is the easiest for analytical treatment. This amounts to the well-known MSE:

$$E_m = e_m' e_m = (b_m X - y_m)' (b_m X - y_m), \quad m = 1, \dots, M \tag{4.7}$$

This criterion is convex in parameters and thus possesses an analytical minimum, resulting from setting the derivatives with regard to parameters b_m equal to zero:

$$\frac{\partial E_m}{\partial b_m} = (b_m X - y_m) X' = 0 \quad m = 1, \dots, M$$

$$b_m X X' = y_m X \qquad\qquad m = 1, \dots, M \tag{4.8}$$

$$b_m = y_m X' (XX')^{-1} \qquad m = 1, \dots, M$$

Organizing the rows of b_m back to matrix B, the result is

$$B = YX'(XX')^{-1} \tag{4.9}$$

This relationship is the well-known solution to the *least squares problem* of *linear regression*.

The factor to the right from Y is a construct known as *pseudoinverse* or *Moore-Penrose inverse* [8,9] of matrix X, usually denoted as X^+. This pseudoinverse is a generalization of the usual matrix inverse. While a matrix inverse is defined only for non-singular square matrices, a pseudoinverse exists for every matrix A. For real-valued matrices, it satisfies several basic relationships:

$$AA^+A = A$$
$$A^+AA^+ = A^+$$
$$(AA^+)' = AA^+ \tag{4.10}$$
$$(A^+A)' = A^+A$$

While a usual matrix inverse has the property $AA^{-1} = A^{-1}A = I$ (its product with the original matrix is an identity matrix), the last two equalities of (4.10) state only that this product is symmetric. However, the first two equalities state that multiplying A by this product leaves A unchanged—as if it were multiplied by the identity matrix. This property can be circumscribed as A^+ being the inverse of A in a subspace defined by A (i.e., the lower-dimensional space corresponding to the rank of matrix A).

The computing definition of the pseudoinverse varies with the properties of matrix A. If A has a full row rank (that is, its rows are not linearly dependent), the definition is

$$A^+ = A'(AA')^{-1} \tag{4.11}$$

This is the case with the least square solution of (4.9). There, it is assumed that the following two conditions hold:

1. Matrix X has a full row rank, that is, the rows of X are linearly independent. In other words, none of the pattern vector elements is a linear combination of other elements across all training examples. This is a practical condition: Such a dependent element would be redundant and may be omitted without information loss.
2. Length N of the input patterns is smaller than the number K of training examples.

The latter condition is essential since it is not automatically satisfied in practice. The length of the input pattern results from available features measured. The number of training examples is determined by the size of the collected data set. Omitting either of both would be information loss. Is the input pattern vector length N (number of rows of X) larger than the number of training examples (columns of X), another definition of pseudoinverse applies. Generally, for a full column rank of matrix A, it is

$$A^+ = (A'A)^{-1}A' \tag{4.12}$$

In our case, this applies for linearly independent input patterns. The optimal mapping matrix B is then, instead of (4.9),

$$B = Y(X'X)^{-1}X' = YX^+ \tag{4.13}$$

As in (4.5), such matrix B is a solution of the equation

$$BX = Y \qquad (4.14)$$

However, with $N > K$, we have the underdetermined case with more variables than equations. Such an equation system has infinitely many solutions for matrix B. The general form of these solutions is

$$B = YX^+ + W\left(I - XX^+\right) \qquad (4.15)$$

with an arbitrary matrix W of size corresponding to matrix B.

Each of these solutions would forecast all output patterns exactly. The pseudoinverse of the form (4.12) selects the solution with the smallest norm, that is, with a tendency of containing small coefficients.

To illustrate the properties of this solution, let us consider a trivial example. Suppose output y is a polynomial function of the input s:

$$
\begin{aligned}
y &= \sum_{j=1}^{N} b_j s^{j-1} \\
&= \begin{bmatrix} b_1 & b_2 & \cdots & b_N \end{bmatrix}
\begin{bmatrix} 1 \\ s \\ \vdots \\ s^{N-1} \end{bmatrix} \qquad (4.16) \\
&= bx
\end{aligned}
$$

with a coefficient vector b and an input vector x consisting of powers of s.

Generating K sample input/output pairs (x_k, y_k) will constitute a training set, that is, matrices X and Y with columns x_k and y_k respectively. In the trivial case of a second-order polynomial, the input data matrix may be

$$X = \begin{bmatrix} 1 & 1 & 1 \\ 0 & 1 & 2 \\ 0 & 1 & 4 \end{bmatrix}$$

and the output data the values of the polynomial function for individual values of s (here: $s \in \{0, 1, 2\}$). With coefficients

$$b = \begin{bmatrix} 1 & 2 & 2 \end{bmatrix}$$

the output pattern matrix (with a single row) would be

$$y = \begin{bmatrix} 1 & 5 & 13 \end{bmatrix}$$

With the fifth-order polynomial coefficient set

$$b = \{ -504 \ 1010 \ -673 \ 189 \ -23 \ 1 \} \qquad (4.17)$$

(generated to produce a predefined set of roots $\{1, 2, 4, 7, 9\}$ that fit into a reasonable plot scale), we receive the function (a polynomial of degree $N - 1 = 5$) depicted in Fig. 4.1.

Fig. 4.1 Fitting a
polynomial function, exactly
determined

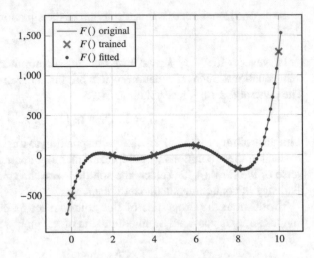

Fig. 4.1 Fitting a
polynomial function, exactly
determined

Fig. 4.2 Underdetermined
case of fitting a polynomial
function

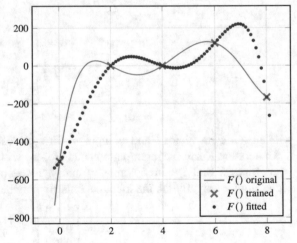

Input vectors, containing the powers of s from 0 to 5, have a length of $N = 6$.
Taking $K = N = 6$ samples, all output can be fitted exactly, that is, all crosses in 4.1
are on the original polynomial function plot. This is the case of neither overdetermination nor underdetermination. Moreover, the mapping vector b (in fact, a single-row
matrix B of preceding formulas) found via (4.5) is identical with the original (4.17).
This is why the fitted polynomial function (solid line) is identical with the original
polynomial function (dotted line) for all intermediary points.

By contrast, reducing the training set to $K = 5$, we receive the underdetermined
case. The mapping by Eq. (4.13) is one of the possible mappings—that with the
smallest norm. This case is shown in Fig. 4.2.

Although the fitted function hits all training set data (denoted by crosses) exactly,
it is not identical with the original function at intermediary points. The minimum
norm property of (4.13) does not guarantee the identity—the original function may
have not been that with a minimum norm. It can only be argued that small norms oc-

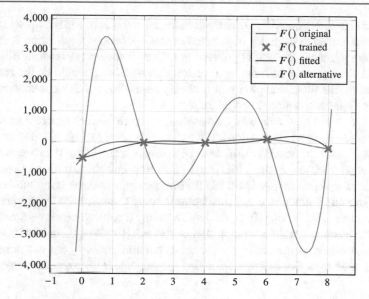

Fig. 4.3 Underdetermined case of fitting a polynomial function, including a non-minimum norm solution

cur more probably than large ones. Nevertheless, resigning to the minimum property may lead to much less acceptable results. One of the infinitely many alternative solutions, received from (4.15), is that of Fig. 4.3 where an alternative solution without minimum norm is included. The deviations of intermediary, untrained points from the original functions are much larger than for the minimum norm solution.

These findings are important for results obtained by methods other than those using the matrix pseudoinverse. In neural network practice, gradient-based algorithms are used. In the underdetermined case, gradient algorithms with good convergence properties (which is not the case for all gradient algorithms, see Chap. 5) would find a zero MSE solution, whose existence is guaranteed in the underdetermined case. However, it will be an arbitrary one, possibly with a large norm. This solution may be very bad at intermediary points. This has motivated gradient methods with additional terms forcing the optimization to prefer small norm solutions by weighted terms (see regularization discussed in Sect. 4.7). But weighting such terms is no guarantee to receive a norm optimum with simultaneously zero MSE.

In the underdetermined case, it is always possible to reach a zero MSE on the training set. If the optimum found is substantially different from zero, it is an indication that the applied optimization method has failed. The solution based on the pseudoinverse selects from the space of infinitely many equivalent solutions those with the minimum norm. Gradient methods do not possess this property—their solution is, without further provisions such as regularization terms, arbitrary. Even with regularization, there is no guarantee that the minimum norm solution is the closest to the "true" mapping—it is only the preferred solution in the sense of regularization theory.

In the underdetermined and exactly determined cases, zero MSE is always attainable (assuming no redundancy of pattern features). A serious implication is that this is the case whatever the output patterns are. Even if they were random (that is, if they had nothing to do with the input patterns), a zero MSE would be the result. In other words, the fitted mapping will probably be useless for any application. This principle is discussed more in detail in Sects. 4.4–4.6.

This fact is not contradicted by occasionally experienced or reported success of such underdetermined applications. Beyond having good luck (and being probably down on this luck the next time), this may be a paradoxical effect of poorly converging optimization methods. At the initial phase of each optimization, the most promising, rough directions for improvement are followed. A poorly converging method may never leave this initial phase and thus optimize only a small, initially promising set of parameters. This is roughly equivalent to having a reduced set of parameters for which the learning problem is not underdetermined. However, there is no guarantee that this principle works under any (or most) circumstances—poor convergence is what its name says. So, it is recommended to avoid underdetermined settings.

4.2   Linear and Nonlinear Mappings

The analyses of the present chapter explicitly apply to linear mappings of the form

$$y = Bx \tag{4.18}$$

The linear case is the mapping class by far the best accessible by analytical methods. The term "linear" is to be made more specific as "linear in parameters":

$$y = Bh(x) \tag{4.19}$$

Function $h(x)$ is a possibly nonlinear function of the input vector x. This function is fixed, i.e., its form is not a part of the parameter fitting process. This encompasses, for example, polynomial mappings if the input data consists of sufficiently many powers of a scalar variable. Such polynomial mappings are used for illustration in the next sections. Another example may be harmonic terms (for periodic mappings being a linear function of *sine* and *cosine* of scalar angles).

It is much more difficult (and sometimes hopeless in practical terms) to analyze general nonlinear mappings parameterized by a vector w of the form

$$y = f(x, w) \tag{4.20}$$

This is why most useful insights into the laws governing nonlinear mappings have been gained with the help of *linearization*. The linearization of a nonlinear function corresponds to the two first terms (the zero and first-order terms) of the well-known Taylor expansion. An inherently nonlinear mapping cannot be made universally linear—the linearization takes place, as does the Taylor expansion, around a certain argument point and is a good approximation only in the more or less close environment of this point.

Assumed continuity and differentiability, nonlinear parameterized mappings can be linearized for any given input pattern column vector x_k around the column parameter vector w_0 as

$$
\begin{aligned}
y &= f(x_k, w) \\
&\approx f(x_k, w_0) + \frac{\partial f(x_k, w_0)}{\partial w}(w - w_0) \\
&= G_k w + (f_k - G_k w_0) \\
&= G_k w + h_k
\end{aligned}
\tag{4.21}
$$

The parameter vector w_0 is arbitrary, but here, it will be chosen to be equal to the optimal parameter vector corresponding to the error minimum. Then, conjectures about the overdetermination or underdetermination will apply to the optimum parametrization.

These approximate equalities for individual pairs of input/output training patterns can be summarized in a single matrix equation

$$
r = \begin{bmatrix} y_1 \\ \vdots \\ y_K \end{bmatrix} - \begin{bmatrix} h_1 \\ \vdots \\ h_K \end{bmatrix} = \begin{bmatrix} G_1 \\ \vdots \\ G_K \end{bmatrix} w = Gw
\tag{4.22}
$$

Here, vector r forecast using parameters w_0 including the shift to the coordinate origin. This equation has, depending on the dimension and the rank of matrix G, the exact or approximate pseudoinverse-based solution for the mapping parameters

$$
w = G^+ r
\tag{4.23}
$$

The rows of matrix G correspond to all output pattern elements to be fitted to. Their number is MK. The columns correspond to individual independent mapping parameters, whose total count is P. (4.22) is a system of MK equations with P unknown variables. The rank of G cannot exceed $\min(MK, P)$. The relation between the rank, MK, and P determines whether there is an exact or only an approximate solution in the sense of least squares and also whether the solution is unique. For a full rank of G equal to $\min(MK, P)$, the relationship between MK and P decides about overdetermination or underdetermination:

- For $P > MK$, the fitting task is underdetermined, having more variables (i.e. mapping parameters) than equations.
- For $P < MK$, the fitting task is overdetermined, having fewer variables than equations.
- For $P = MK$, the fitting task is exactly determined.

This is an analogy to the underdetermination and overdetermination concepts of linear mappings. To establish the equivalence relationship to a linear mapping $y = f(x, B) = Bx$, let us consider (4.21) in this special case. The following derivations require the use of not very pictorial concept of *Kronecker product* defined below. This is why they are admittedly not easy to follow and can be skipped if the reader is not interested in the detailed algebraic argument.

Since the model is linear, Eq. (4.21) is exact and not approximated. Around $w_0 = 0$, it can be written as:

$$\frac{\partial f_m (x_k)}{\partial b_{m'n}} = \begin{cases} x_{nk}, & m' = m \\ 0, & m' \neq m \end{cases} \tag{4.24}$$

The selection of $w_0 = 0$ is arbitrary but the results would be identical for any other parameter point due to the linearity of the mapping. For $w_0 = 0$, the term h_k is zero and the equation for the m-th element of k-th output pattern vector y_k is

$$y_{mk} = \sum_{n=1}^{N} \frac{\partial f_m (x_k)}{\partial b_{mn}} b_{mn} = \sum_{n=1}^{N} x_{nk} b_{mn} = x_k b_m \tag{4.25}$$

The matrix G_k of partial derivatives for the k-th pattern is

$$G_k = \begin{bmatrix} x_k' b_1' \\ \vdots \\ x_k' b_M' \end{bmatrix} \tag{4.26}$$

In the following derivation, the Kronecker product $(X' \otimes I_M)$ is helpful. It is defined in the following way. If A is an $m \times n$ matrix and B is a $p \times q$ matrix, then the Kronecker product $C = A \otimes B$ is the $pm \times qn$ block matrix

$$C = [a_{ij} B] = \begin{bmatrix} a_{11} B & \cdots & a_{1n} B \\ \vdots & \ddots & \vdots \\ a_{m1} B & \cdots & a_{mn} B \end{bmatrix} \tag{4.27}$$

The column parameter vector w can be organized as a transpose of concatenation of row parameter vectors b_m

$$w = [b_1 \ldots b_M]' \tag{4.28}$$

The matrix G_k of partial derivatives (4.26) can be expressed as

$$G_k = (I_M \otimes x_k') \tag{4.29}$$

with I_M being the unit matrix of dimension M.

The matrix equation for all patterns of the training set is then

$$\begin{aligned} Y_{\text{vec}} &= \begin{bmatrix} y_1 \\ \vdots \\ y_K \end{bmatrix} \\ &= G w \\ &= \begin{bmatrix} X' & 0_N & \cdots & 0_N \\ 0_N & X' & \cdots & 0_N \\ \vdots & \vdots & \ddots & \vdots \\ 0_N & 0_N & \cdots & X' \end{bmatrix} [b_1 \cdots b_M]' \\ &= (I_M \otimes X') B_{\text{vec}} \end{aligned} \tag{4.30}$$

with 0_N being a zero matrix of the size of X'. The vectors Y_{vec} and B_{vec} are vector variants of matrices Y and B, constructed by concatenating the columns of matrix Y and transposed rows of matrix B.

The pseudoinverse of Kronecker product $(A \otimes B)$ is $(A^+ \otimes B)$, in our case $(I_M \otimes X'^+)$. Consequently, the solution is

$$B_{\text{vec}} = [b_1 \cdots b_M]' = (I_M \otimes X'^+) Y_{\text{vec}} \tag{4.31}$$

This is equivalent to $B = YX^+$. The Kronecker product has the rank corresponding to the product of both factors' ranks, $M \times \min(N, K)$ in our case.

After having swallowed the Kronecker product formulas, we can state that the consequence is the following: Instead of the relationship between N and K, the existence and uniqueness of the mapping parameter vector are determined by equivalent conditions for MN (the length of parameter vector w) and MK (the length of vector Y_{vec} corresponding to the size of output data matrix Y). The former product corresponds to the number of mapping parameters P defined above. Applying the conditions with the pair N and K and those with the pair MN and MK lead to equivalent results concerning the uniqueness of the mapping parameter vector, whether exact or optimally approximated.

Let us now go back to the general nonlinear mappings. The conjectures of Sect. 4.1 as well as of the following Sects. 4.3–4.7 are valid in the corresponding manner. Their analysis can be performed for the environment of any initial parameter vector w_0. Of course, the rank of matrix G, and thus, the results of the analysis can theoretically be different in environments of different vectors w_0. This will particularly be the case with non-smooth activation functions such as rectified linear units as in (3.5). If some input patterns lead to zero activation of this node type, the partial derivative of incoming weights will be zero, and the rank of the partial derivative matrix G can be reduced. The rank variability will be much scarcer for sigmoid activation functions (3.2) and (3.3).

However, even if the ranks of matrices G for different parameter points w_0 are identical, they may be generally reduced by the neural network topology and its inherent invariances (e.g., permutations of hidden units). In other words, the case of full rank equal to $\min(MK, P)$ is mostly not encountered.

An example is a simple network of two linear layers with weight matrices B and C (as in Fig. 4.4)

$$\begin{aligned} y &= Bz \\ z &= Cx \end{aligned} \tag{4.32}$$

corresponding to

$$y = BCx \tag{4.33}$$

With the hidden vector z of width J, there is a $(M \times J)$-matrix B and a $(J \times N)$-matrix C. The number of mapping parameters is $MJ + JN$. Such architecture makes sense only if $J < M$ and $J < N$, so that BC is not full rank. Otherwise, BC would be equivalent to a single arbitrary weight matrix $W = BC$. However, with $J < M$, $M - J$ output elements have to be linearly dependent on the remaining J outputs

Fig. 4.4 Two-layer linear
network

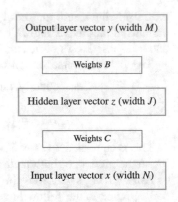

Output layer vector y (width M)

Weights B

Hidden layer vector z (width J)

Weights C

Input layer vector x (width N)

to be able to be forecast exactly. So, only J^2 elements B are independent, free parameters while the remaining $(M - J)J = MJ - J^2$ are dependent. Additionally, the mapping (4.32) is equivalent to

$$y = BHz$$
$$z = H^{-1}Cx \qquad (4.34)$$

with an arbitrary non-singular $(J \times J)$-matrix H. This arbitrary matrix reduces the degrees of freedom by J^2. So, instead of $MJ + JN$, the rank of the matrix of partial derivatives G from (4.22) is bounded by $J^2 + JN - J^2 = JN$, although its column number remains $MJ + JN$. This has consequences for the existence of the exact solution.

For more complex neural network topologies, the rank loss may be larger and sometimes nontrivial to determine.

In any case, the inequality $P \le MK$ implies the overdetermined case—there are too few mapping parameters to be fitted to the set of MK given constraints. This is true whenever the rank of the output data matrix Y is full. This condition is usually satisfied unless there are either redundant training examples (being linear combinations of other examples) or redundant output pattern elements (being linear combinations of other pattern elements).

With this provision, the considerations of the following Sects. 4.3–4.7 concerning overdetermined and underdetermined cases apply correspondingly, if the lack of precise information about the number of genuinely independent mapping parameters is considered.

The cases of overdetermination, underdetermination, and exact determination in nonlinear systems are analogous to those of linear systems. They can be investigated through linearization at a certain parameter point, particularly that corresponding to the error minimum. The key numbers are the number of independent mapping parameters P and the number of fitting equations MK, with the width of output pattern M and the number of training examples K. With a full-rank matrix of partial derivatives G, the overdetermined case would be characterized by $P < MK$ and the underdetermined one by $P > MK$.

Additionally, there is a possible (and frequent) rank loss by invariances inherent in neural network topologies. This rank loss will lead to shifting the boundary $P = MK$ toward underdetermined case (i.e., seemingly underdetermined cases may turn out to be overdetermined, leading to a failure of exact fitting).

An analogical treatment of linear mappings with $P = MN$ would amount to the comparison between MN and MK. They can be analyzed more easily, comparing N and K (which are the two dimensions of the input matrix X of the training set).

4.3 Overdetermined Case with Noise

Let us now have a closer look at overdetermined cases with optimum linear mapping according to (4.9). It is usually assumed that the application data has been generated by a defined process. This means that there is some "true" mapping from input x to output y. In reality, this mapping is distorted in various ways:

- There may be noise added to the measurements of both input and output.
- The true mapping is a function that can be approximated by the given parameterized mapping (e.g., linear regression or neural network) only with limited accuracy.
- The process is not stationary—the true mapping changes during data collection.

Although these factors differ in many details, they have a common consequence: an unexplained residue will remain, resulting in a residual error.

Let us once more consider a mapping corresponding to a polynomial function. It is a polynomial of the fifth order so that the output y depends on the zeroth to the fifth powers of a scalar variable. Let us observe its approximation by a polynomial of the fourth order, so that the training set matrix X consists of zeroth to the fourth power of variable s, that is, columns of length $N = 5$. Consequently, no accurate fit can be reached.

In the underdetermined and the exactly determined cases, using the mapping (4.9), the fit to the training set is perfect. This is shown in Fig. 4.5 ($K = 4$, $K < N$) and Fig. 4.6 ($K = N = 5$). However, for $K > N$, the task becomes overdetermined. Already with $K = 6$ (see Fig. 4.7), approximation errors are inevitable. The nature of the approximation is even better illustrated by Fig. 4.8 with $K = 25$. The fit to the training set symbolized by crosses corresponds to the best approximation of the fifth-order polynomial by a fourth-order polynomial.

In the presence of noise or if the true mapping cannot be exactly approximated by the given mapping (regression or neural network), a perfect fit to the training set is not feasible. If the number of training samples is such that the task is overdetermined, approximation error on the training set is inevitable.

How far a mapping can be approximated by another mapping from a class the true mapping does not belong to, can be answered only for individual cases. If the noise

Fig. 4.5 Underdetermined
case of polynomial
approximation

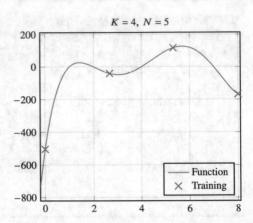

Fig. 4.6 Exactly determined
case of polynomial
approximation

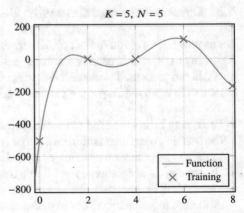

Fig. 4.7 Overdetermined
case of polynomial
approximation with $K = 6$

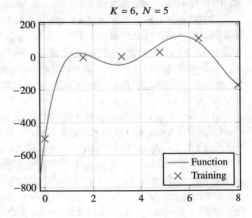

Fig. 4.8 Overdetermined case of polynomial approximation with $K = 25$

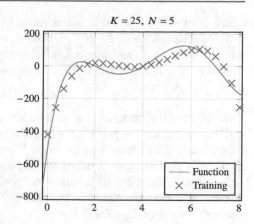

is of Gaussian type, that is, additive, independent between the training samples and not correlated with the output, its influence can be made more precise. In the linear case, the forecasted output has the form

$$y = Bx + \varepsilon = \eta + \varepsilon \qquad (4.35)$$

It consists of the deterministic part $\eta = Bx$ and the noise part ε. Both parts are supposed to be independent. For simplicity of presentation, we can assume that the deterministic component is satisfied exactly and to investigate only the "forecast" of the noise ε. This will make the formulas more transparent and comprehensible. The word forecast has been set to parentheses since it is an undesirable forecast. We know that a random component of Gaussian noise type cannot be forecasted. Training examples are used to find out what the deterministic component η may be. By contrast, random noise is to be neglected as far as possible. Neglecting this is the opposite of fitting the model to the noise. The goal of the following argument is to show to which extent this undesirable fitting takes place, depending on the training set size.

The forecast deviations for the training set can be organized in the matrix

$$E = Y - BX \qquad (4.36)$$

with B according to Eqs. (4.5), (4.9) and 4.13, respectively.

For the exactly determined case, Eq. (4.36) is

$$E = Y - YX^{-1}X = Y - Y = 0 \qquad (4.37)$$

that is, the deviation is zero. In other words, the parameters are completely fitted to the noise—a undesirable result.

For the underdetermined matrix B, the deviation will vanish, too:

$$E = Y - BX = Y - Y(X'X)^{-1}X'X = 0 \qquad (4.38)$$

However, the overdetermined case amounts to

$$E = Y - BX$$
$$= Y - YX'(XX')^{-1}X \qquad (4.39)$$
$$= Y\left(I - X'(XX')^{-1}X\right)$$

which cannot be simplified in an immediately obvious way.

To understand the nature of this solution, an important concept of linear algebra must be introduced, the *Singuar Value Decomposition* (SVD).

Every $(N \times K)$-matrix A can be decomposed into the following three factors:

$$A = UDV'$$
(4.40)

In the so-called compact form (where all columns and rows multiplied by zeros and thus not affect the product omitted), the matrices U and V have orthonormal columns and their dimensions are $H \times N$ and $H \times K$, respectively. The number $H, H \leq N$ and $H \leq K$, is the rank of matrix A. The matrix D is of size $H \times H$ and diagonal. The diagonal elements are positive real numbers usually ordered in magnitude descending from the upper left corner to the lower right. These diagonal elements are called *Singular Values*.

The orthogonality of columns of U and V allows the representation as

$$A = \sum_{i=1}^{H} u_i d_i v_i'$$
(4.41)

called *spectral decomposition* of matrix A. Here, u_i and v_i are column vectors and d_i are scalar singular values from the diagonal matrix D. This representation is useful in pointing out important (those with larger singular values) and unimportant, sometimes negligible, components of matrix A (those with small singular values).

From many useful properties of singular value decomposition, we will use its simple expression for matrix pseudoinverse. It is in the following

$$A^+ = VD^+U'$$
(4.42)

with diagonal matrix D^+ having nonzero diagonal elements $\frac{1}{d_i}$ reciprocal to those of D. The zero diagonal elements of D remain zero.

The product of matrix A with its pseudoinverse A^+ is

$$A^+A = VD^+U'UDV' = VV' = \sum_{i=1}^{H} v_i v_i'$$
(4.43)

This expression is known as the *projection operator* to a lower-dimensional linear space defined by the vectors v_i. For example, it would project points in three-dimensional space to a plane defined by two vectors v_i, or points in a two-dimensional plane to a straight line defined by a single vector.

Applied to the deviation between forecast and measured output values in the overdetermined case (4.39), we receive

$$E = Y\left(I - X^+X\right) = Y\left(I - VV'\right) = Y\left(I - \sum_{i=1}^{H} v_i v_i'\right)$$
(4.44)

For $H = K$, the projection would take place to the complete space of input patterns vectors—the expression in parentheses would be zero. In the overdetermined case, the rank H of the input data matrix X is less than K. In the usual case of mutually independent measured features (rows of X), H is equal to the input vector width N.

Generally, the forecast deviation would depend on the values of the output matrix Y—some output pattern vectors coincide with the projection space, others do not. The former are "easier" to forecast.

However, the relationship (4.44) sheds light on the attainable MSE in the presence of noise. If the output patterns consisted only of the white (i.e., Gaussian) noise of unit variance, they could be perfectly fitted for $K = N$ (the exactly determined case). Of course, this "fit" would be an illusion—only those random values of Y that are present in the training set would be able to be exactly forecast. For $K > N$ (the overdetermined case), the K-dimensional space of white noise would be projected to the N-dimensional subspace. In other words, only a subset of N random values can be fitted to, the remaining $K - N$ cannot. From the mean squared value of noise components (which is assumed to be equal to one in our consideration), the fraction of $\frac{N}{K}$ can be forecast and $\frac{K-N}{K}$ cannot. With growing K, the latter fraction converges to unity. The MSE attainable for K training examples with white noise of standard deviation σ and variance σ^2 is

$$\frac{K - N}{K}\sigma^2 \qquad (4.45)$$

The size of variance is usually unknown. It can be estimated only with the knowledge of the true model which, in turn, can only be determined with the help of the training set and even then only if the training set is of sufficient size. The true magnitude of the term (4.45) will not be known a priori and cannot be considered during the training. It is only the fractions $\frac{N}{K}$ and $\frac{K-N}{K}$ themselves that are informative. They tell us how much of the noise has been made invisible by fitting because of a too small training set.

This can be confirmed by simulations. In Fig. 4.9, the MSEs for training sets consisting only of unit variance Gaussian noise, with various training set sizes $K > N$, are shown. The data are powers of scalar variable s as for polynomial approximations used above, with $N = 6$.

Fig. 4.9 MSE when forecasting white noise of unit variance in the training set

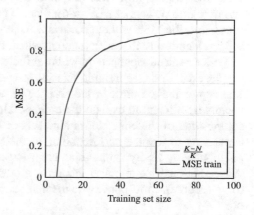

Fig. 4.10 Polynomial
function with sufficient
input: MSE on training set
with and without additive
noise

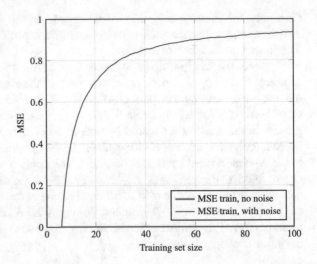

The attainable MSE develops consistently with the ratio $\frac{K-N}{K}$.

If the output training vectors contain both true patterns and an additive noise which is not correlated with these noise-free patterns, the MSE is the sum of

- approximation error of true patterns; and
- the unfitted part of the noise variance.

If the patterns can be, except for noise, completely explained with help if input patterns and so their forecast error is zero, the MSE corresponds to the unfitted noise variance as in Fig. 4.9. This is the case, for example, if the output patterns are polynomial values of an order completely covered by the input patterns. In Fig. 4.10, the output are the fifth-order polynomial values and the input patterns consist of zeroth to fifth powers of s ($N = 6$). Since a perfect fit to output polynomial values is possible, the MSE without noise is zero. With additive unit variance white noise, MSE is identical to that of noise component, the fraction $\frac{K-N}{K}$.

With a reduced input pattern width, sufficient only for the fourth-order polynomials, the exact fit on the training set is no longer possible. This is shown in Fig. 4.11. The approximation error in the absence of noise (blue curve) is due to the input vectors insufficient to explain the outputs. The shape of this curve depends on the composition of the concrete training sets of given sizes. With a growing K, it is converging to a value corresponding to the best approximation of a higher-order polynomial by a polynomial of an order lower by one, on a given argument range. With additional white noise, this error grows (red curve). The difference of both (orange curve), corresponding to the noise component, follows the theoretical value of $\frac{K-N}{K}$.

The principles governing the fit to the training set in the overdetermined case can be summarized in the following way:

Fig. 4.11 Polynomial function with reduced input: MSE on training set with and without additive noise

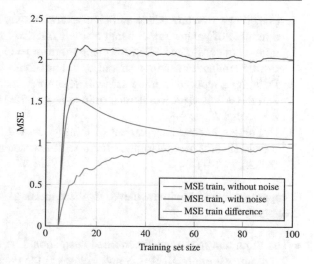

The best fit in the sense of MSE encompasses both true output data patterns and noise, although fitting to the latter is not desirable. Fitting to noise takes place to a larger extent in the proximity of exactly determined case, that is, for the number of training examples K only moderately exceeding the training pattern vector width N. For large K, this effect diminishes— the fit approaches the best possible approximation of the true model, given the individual input pattern.

Nonlinear mappings in Sect. 4.2 are governed by similar laws, with the mentioned problems to determine the number of degrees of freedom that can be lower than the number of parameters. The proportion of noise shadowed by undesirable fitting is $\frac{MK}{P^+}$, with $P^+ \leq P$ being the number of genuinely free parameters. For nonlinear mappings, there are some additional problems making the fitting with exclusion of noise difficult:

- Fitting to nonlinear dependencies is influenced by the type of nonlinearity. Some nonlinearities are difficult to represent by means of other nonlinearity types.
- While error minima for linear mappings are known in an analytical form, those for nonlinear ones can only be found by numerical optimization. Nonlinear optimization is challenging task, depending on the mapping type, as argued in Chap. 5. If the algorithm used does not reach the genuine error minimum, it cannot be expected to be able to control the undesirable fitting to noise according to the quotient given above.

A possible procedure for checking the interaction between the optimization algorithm and the degree of overdetermination may be the following. For this, it is useful to define the overdetermination factor $f = \frac{MK}{P}$.

1. Selecting a structure with a sufficient extent of overdetermination, that is, with a number of parameters P being a small fraction (say, maximum 25%) of the number of constraints MK for a given training set of size K. In other words, the overdetermination factor f should be at least four.
2. Defining a reduced training set with K_u representatively drawn samples such that the task is clearly underdetermined (e.g., with overdetermination factor of 0.5 so that $K_u = \frac{1}{2}\frac{P}{M}$).
3. Defining a intermediary variant of training set with K_o samples such that the task is slightly overdetermined (e.g., with overdetermination factor f_o two so that $K_o = 2\frac{P}{M}$).

Comparing the error minima achieved for these training sets will deliver the following insights:

- The minimum in the overdetermined case should be close to zero (relatively to the other two minima). If it is not, either (1) the data are governed by extreme nonlinearities that cannot be modeled or (2) the optimization algorithm shows poor convergence, not being able to find the correct solution equal to zero.
- The minima of the full training set fitting and the slightly overdetermined set fitting should be approximately proportional to $1 - \frac{1}{f}$ and $1 - \frac{1}{f_o}$. If not, the number of genuinely free parameters P^+ is substantially smaller than P.

For nonlinear models such as those based on neural networks with P mapping parameters (e.g., network weights and biases), an analogical law can be expected for the number of constraints equal to MK moderately exceeding the number of independent parameters P.

It has to be stressed that the development of MSE according to Figs. 4.9 and 4.11 takes place with growing training set size and fixed parameter set. In the inverse situation with a fixed training set and a growing number of parameters, the white noise component of MSE may be supplemented by the possible deviation caused by an insufficiently expressive model (e.g., a polynomial model of lower order than the real system on which the data have been measured).

4.4 Noise and Generalization

The ultimate goal of building a model representing a mapping, whether with classical linear methods or with the help of neural networks, is a good prediction performance for cases encountered after fitting the parameters has been accomplished. The training set is only a means for determining the parameters. Accordingly, high performance on the training set is useful only if it goes hand in hand with a correspondingly high performance on some test set representing the "rest of the world" not included in the training set.

The correspondence of good performance on both the training and the test set is called *generalization*. It should confirm the model being sufficiently general to be valid beyond the training set. There is a large corpus of theory around generalization, leading to a diversity of opinions how to reach the best results in practice. On the very rough scale, two basic approaches can be discerned. The first one consists in trying to reach the best possible results (i.e., the least forecast error) for the training set, expecting that this will also lead to good results on the test set. The main advantage of this approach is that the performance on the training set can be deterministically measured and, in some cases, reaching the optimum supervised. An important prerequisite for the success is an appropriate ratio between the number of free parameters and the size of the training set to exclude undesirable fitting to the noise, as discussed in Sects. 4.1 and 4.2.

The other basic approach is the statistical one. Its point of departure is the expected performance on the whole statistical population, that is, also on the test set as a random sample from this population. Also the training set is viewed as a random sample. Essential ideas are based on the decomposition of the square forecast error to the square of forecast bias and the variance of the forecast. (Forecast bias is the deviation of the expected forecast mean from the true mean.) It has been proven that under some conditions, the forecast error on the statistical population may be smaller if a certain amount of nonzero bias is accepted. This is the main difference to the former approach where minimizing the error on the training set leads to a zero bias.

The conditions for minimum forecast error with nonzero bias are difficult to prove in practice and are discussed in Sect. 4.7.1. Following this approach amounts to giving up the precision that can be directly monitored (training set performance) for statistically founded hope for being better off in generalization (something that cannot be directly observed, except with drawing sufficiently large validation samples).

Because of this complex decision situation, we will mostly follow the former approach, unconditional optimization on the training set. Approaches with nonzero bias will be discussed in Sect. 4.7.

We start with a glance at the generalization in the underdetermined and the overdetermined cases.

In the underdetermined case of too few training examples for unambiguously determining the model parameters, the chances for good generalization are particularly bad. It has been shown in Sect. 4.1 that there are infinitely many parameter solutions exactly consistent with the training data. The solution based on the input data matrix pseudoinverse is outstanding in its property of having the minimum norm. However, it has been argued that the minimum norm is only a kind of "most probable" case, without any guarantee of being close to the true process generating the data. The underdetermined case will be addressed in more detail in Sect. 4.5.

The overdetermined case is a more favorable configuration. The parameter optimum concerning the given training data set is unambiguous. Nevertheless, with the omnipresent noise, some possibly unexpected distortions are to be considered.

For a given training set input pattern x, the deviation of the forecast $f(x)$ from the correct output pattern y is fixed after the fitting process has been accomplished. By contrast, for a test set that is a random sample from the statistical universe of the application, the deviation is a stochastic variable that can only be described in statistical terms. It can be decomposed into three parts [7]:

$$E\left[(y - f(x))^2\right] = E\left[\epsilon^2\right] + (y - E[f(x)])^2 + E\left[(f(x) - E[f(x)])^2\right] \quad (4.46)$$

The terms at the right side correspond, respectively, to

(a) inevitable error cause by random noise ϵ;
(b) the square of the systematic deviation of the estimate, that is, the mean bias; and
(c) the variance of the estimate.

The least square procedure delivers unbiased estimates with (b) equal to zero.

As mentioned above, procedures with biased estimates are also used. The motivation for this is the chance to reduce the total MSE of the estimate in comparison with the least squares, which is possible only with a careful choice of tuning parameters. An example of such procedure is the *Ridge Regression*, briefly discussed in Sect. 4.7.1.

Even if the fitting procedure is such that the bias is zero, there are two sources of fluctuations in the presence of noise (for both a linear mapping or a neural network with given connectivity and activation functions):

1. The noise present in the test data will inevitably lead to deviations between the forecast and the measured output values. Even a perfect model of the true process would exhibit these deviations.
2. The noise present in the training data affects the model parameters in the way discussed in Sect. 4.3. The model parameters are fitted to the noise, if the number of constraints exceeds the number of free parameters only moderately. For linear mappings, the condition concerns the training set size K and its excess over the input pattern vector width N. This uncertainty concerning the model parameters will be reflected by additional imprecisions on the test set. This corresponds to the component (c) mentioned above.

In the case of uncorrelated white noise, both effects are additive.

For linear mappings, some analytical results are available. The latter source of fluctuations is the easiest to analyze for individual rows of the output pattern matrix. According to Eq. (4.35), the fitted forecast on the training set is $y = \eta + \varepsilon$, that is, it has two components: the deterministic model output η and the random noise ε. The former component can theoretically be fitted exactly. The latter component is, if it is white random noise, uncorrelated with the deterministic component. To simplify the formulas, we will consider only the second component and its influence on the corresponding part of the regression vector b. With such y equal to white noise with unit variance, the least squares regression vector is

$$b = yX'(XX')^{-1} \tag{4.47}$$

Its expected value is zero (which is the best forecast of the white noise), and its variance is

$$\begin{aligned} E\left[b'b\right] &= (XX')^{-1} X E\left[y'y\right] X'(XX')^{-1} \\ &= (XX')^{-1} XX'(XX')^{-1} \\ &= (XX')^{-1} \end{aligned} \tag{4.48}$$

The product XX' is matrix G with elements

$$g_{ij} = x_i x'_j = \sum_{k=1}^{K} x_{ik} x_{jk} \tag{4.49}$$

If the training examples are viewed as samples from the population of possible input patterns, g_{ij} equals to expected value of the product of i-th and j-th pattern element, that is, the second moment of this element pair, multiplied by the number of training examples. The multiplication of the mean by the number K results in the sum in (4.49), which can thus be circumscribed with the help of expected values as

$$g_{ij} = K E_x\left[x_i x_j\right] \tag{4.50}$$

This expected value is not computed over the noise distribution but over the distribution of input features x in the statistical population of the application and is denoted by E_x to make this distinction. In matrix notation, it is

$$XX' = G = K E_x\left[xx'\right] \tag{4.51}$$

with pattern vectors x.

The variance of the regression coefficients corresponds to the inverse of this matrix:

$$E\left[b'b\right] = (XX')^{-1} = G^{-1} = \frac{1}{K} E_x\left[xx'\right]^{-1} \tag{4.52}$$

The inverse of the expected second moment matrix is constant for given statistical properties of the problem. Through the term $\frac{1}{K}$, the variance of the least square coefficients decreases with a growing number of training examples K. This is important since the test set performance directly benefits from this decrease. With non-unit noise variance of size σ^2, Eq. (4.52) is to be multiplied by σ^2. The variance of the model coefficients is

$$E\left[b'b\right] = \frac{\sigma^2}{K} E\left[xx'\right]^{-1} \tag{4.53}$$

The noise variance σ^2 is usually unknown. However, even without its knowledge, the usefulness of Eq. (4.53) consists in showing the diminishing variability of the model coefficients with growing size of the training set. The value of σ^2 is only a constant that rescales this variability dependence.

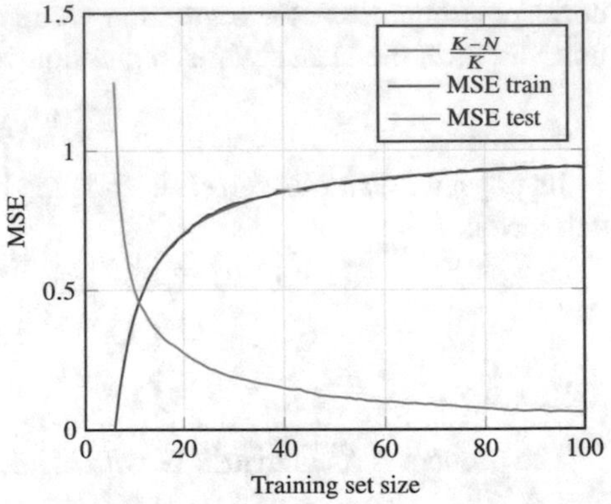

Fig. 4.12 Training MSE
with noise and resulting test
MSE without noise

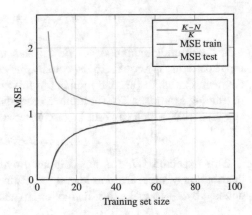

Fig. 4.13 Training MSE
with noise and resulting test
MSE including test set noise

This component of test set error caused by the training set noise is shown
in Fig. 4.12. The test set noise is neglected in this figure, to show the influence of
training set noise alone. It is obvious that the component of the test set error caused
by the training set noise decreases with the training set size, although the training
set error (including noise) grows. (We know that the training set error grows because
the undesirable fit to the training set noise tends to vanish with the training set size.)

Of course, the assumption of no test set noise is not realistic—this simplification
has been shown only for the sake of illustration. In reality, the test set will also suffer
from the noise included in its own patterns (here: white noise with unit variance). The
ultimate result is as depicted in Fig. 4.13. The test set error consists of the component
shown in Fig. 4.12 decreasing with the training set size and the constant level of the
own noise of the test set. With growing training set size, both the training and the
test set errors asymptotically converge to a level given by the noise.

With deviations caused by the missing approximation capability of the model, the
test MSE will further grow. For example, in the case of polynomial approximation
of insufficient order (the fourth-order approximation of the fifth-order polynomial
dependence), even a noise-free training set would produce a deviation. This deviation
will further grow through distortions of model parameters caused by training set noise
(see Fig. 4.14) by a fixed amount corresponding to the noise variance.

Fig. 4.14 The case of insufficient approximation capability of the model: test MSE (noise excluded) resulting from a noise-free and a noisy training set

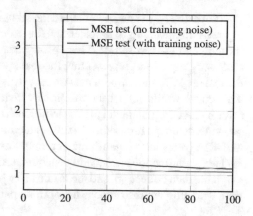

In practice, noise components usually cannot be directly quantified. In particular, it is impossible to assess the sufficiency of the approximation capability of given mapping (linear or neural network) since the true mapping is not known. The extent of measurement noise (whose simplest example is the white noise) is equally unknown. However, at least the qualitative effect of training set size can be kept in mind:

Training set size that is close to the exact determination of mapping parameters (i.e., where the number K of training examples is only moderately larger than the width N of the input pattern vectors) leads to the undesirable fit to the training set noise. This fit is at the expense of the test set performance. This effect is added to the noise of test examples themselves. Whenever possible, it is advisable for the number of training examples to be a sufficiently large multiple of the input pattern vector width.

The gap between the training set and the test set errors converges to zero with growing training set size (if the noise is random and both sets are generated by the same process).

For nonlinear mappings discussed in Sect. 4.2, the modified rule is:

For nonlinear models such as those based on neural networks with P mapping parameters and MK constraints, the overdetermination factor can be defined as $f = \frac{MK}{P}$. A too small (close to unity) overdetermination factor leads to the undesirable fit to the noise. A direct consequence is a poor test set performance.

The convergence of training set and test set MSE shown in Fig. 4.13 refers to the case of fixed model parameter set and growing training set, leading to the growing degree of overdetermination. The training and the test set converge to a horizontal line corresponding to the level of white noise.

Alternatively, the degree of overdetermination can increase through a fixed training set and shrinking parameter set (i.e., simplifying the model). Then, both the training and the test set MSE may grow because of insufficiently expressive model that is not capable of representing the real system. In this situation, the training and the test set MSE plots would not converge to a horizontal line but to a growing curve. This curve would reflect the extent to which there is an inevitable approximation error by insufficient model parametrization. The difference between both reasons for growing determination ratio f is illustrated by hypothetical examples in Figs. 4.15 and 4.16 (plots with logarithmic horizontal axis for better clarity).

This phenomenon is frequently encountered in practice: The training set is fixed by available data collection, and the user cannot but vary the model and its parametrization. The problem is to choose the optimal parameter set. Too many parameters lead

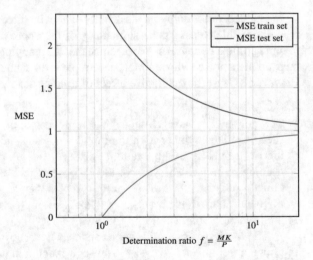

Fig. 4.15 Training and test set MSE with determination ratio f growing through increasing training set size

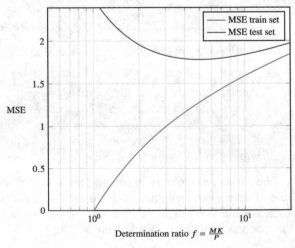

Fig. 4.16 Training and test set MSE with determination ratio f growing through decreasing number of parameters

to a small determination ratio f with the consequence of poor generalization. Too few parameters may be unable to capture the particularities of the system represented by the data. This is a key challenge encountered in all applications.

The model choice is a problem widely discussed in classical statistics and proposals for its solution consist in many computationally expensive trials. For model classes with complex architectures such as neural networks, there is no universal and simultaneously practical solution. This book cannot but emphasize the presence and the difficulty of this problem, without offering generally valid recommendations. Anyway, it may be useful information for the user that the performance and generalization will qualitatively behave as depicted in Fig. 4.16. A partial, but important aspect is addressed in Sect. 4.9.

4.5   Generalization in the Underdetermined Case

The underdetermined case is loaded with a problem that is, in principle, insurmountable as long as zero-bias solution is used. As explained in Sect. 4.1, fitting the parameters to the training set is always possible with a zero MSE but this fit is ambiguous. In other words, there is always a possibility to hit the "true" model. This substantially reduces the expectations for generalization. The regularization approach discussed in Sect. 4.7 has similar properties as the minimum norm principle of pseudoinverse: it may or may not hit the true relationship, depending on whether the prerequisite is satisfied. This prerequisite is the credibility of minimum norm assumption for the pseudoinverse and of prior distribution of parameters in the Bayesian regularization.

A practical and frequently practiced (although not always working) method is the following: to improve generalization, no matter if the problem is over- or underdetermined, it is made use of, in addition to the training and the test sets, a *validation set*. The role of this set is to assess the generalization performance and using this information for model selection. The problem is that such model selection with help of the validation set makes this set to a part of the training set—it is simply "another method" of model building. To understand this, it can be imagined that a simple model building method may be a random search under all possible models and selecting the one with the best performance on this set. If using a validation set, the set used for selection comprises both the training and the validation set.

If the validation set is gained by splitting off some data from the training set, the danger of harmful fitting to training set noise is growing. A fatal mistake would be to reduce the training set as much as to cross the overdetermination boundary toward the underdetermined case. It cannot be expected that under underdetermination conditions, the model ambiguity can be surmounted by testing on a validation set. The reason for this is that there is an infinite number of possible "optimal" model candidates while only a finite number can be realistically tested on the verification set. So, using a validation set is recommendable only if there remains enough data in the training set so that the degree of overdetermination (ratio of training samples to constraints) is sufficient.

Anyway, there is no guarantee that the performance on the test set (which characterizes the true generalization if this set is sufficiently large) is equally good as that on the validation set. The factors determining the difference between the training set performance on one hand and the test set performance on the other hand, are the same: nonstationary data generating processes, random noise and nonlinearities of unknown form.

Nevertheless, a validation set can be viewed as a helpful step in the right direction if it is used to select between a moderate number of distinct model variants. Such variants can be the polynomial model order or different nonlinearity types. Using verification sets for cross-validation is discussed in Sect. 4.8.

In a genuinely underdetermined situation, some useful considerations can be made. The underlying principle consists of making the underdetermined case overdetermined. The means to do this is by reducing the set of independent parameters.

The extent to which this reduction is helpful can be determined with the help of the statements in Sect. 4.3. It has been observed in the overdetermined case that the degree of overdetermination (characterized by the ratio $\frac{K}{N}$ for linear problems and $\frac{MK}{P}$ for nonlinear ones) affects the undesirable fitting of the model to the noise. The larger this ratio, the smaller the adverse effect of this fitting to the test set. In other words, it is advantageous, from the point of view of undesirable fitting to noise, if the rank of the data matrix N is relatively small regarding the data set size K.

The task is to introduce some degree of overdetermination in the underdetermined case. To illustrate this paradoxically sounding idea, let us observe the rank of the mapping matrix B in the linear mapping case. This matrix is the product of the output training data matrix Y of dimension $M \times K$ and the pseudoinverse of the input data matrix X of dimension $N \times K$. The maximum rank of Y is thereafter $\min(M, K)$ and that of X as well as of its pseudoinverse is $\min(N, K)$. Since the rank of a matrix product is less or equal to the ranks of the factors, the maximum rank of B is $\min(M, N, K)$. In the most practical cases, the rank will be exactly $\min(M, N, K)$. The definition of the underdetermined case with $K < N$ allows this term to be simplified to

$$H = \min(M, K) \tag{4.54}$$

For $M < K$, it is obvious that there is little space for manipulating the rank. In particular, for a scalar output with $M = 1$, the number of mapping parameters (equal to N) cannot be reduced if the influence of every of the N input vector elements is to be weighted to reflect its relevance for the output.

However, for $M \geq K$, there is at least some potential for reducing the rank from K to some lower value that may hopefully diminish the effect of fitting the mapping parameters to the random noise.

Let us now apply SVD of (4.40) to the mapping matrix B:

$$B = UDV' \tag{4.55}$$

The diagonal matrix D contains singular values ordered by their size. Deleting singular values that are smaller than a certain threshold is a widespread method to approximate the matrix B by a lower rank matrix with minimum precision loss. (This threshold results from the specification of the maximum the matrix norm of

Fig. 4.17 Linear two-layer network with a *bottleneck* hidden layer

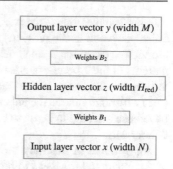

the difference between the original and reduced matrix.) More exactly, deleting the smallest $H - H_{red}$ out of the total of H singular values results in matrix

$$B_{red} = U D_{red} V'$$ (4.56)

which is the closest approximation with rank H_{red} of matrix B in terms of the matrix norm of the difference $B - B_{red}$. As a result of the properties of matrix norm, the mapping $B_{red}x$ is the best approximation of the mapping Bx. Concretely, the difference

$$y - y_{red} = Bx - B_{red}x$$ (4.57)

has the minimum upper bound of its norm for all vectors x with a unit norm.

According to (4.56), matrix B can be written as a product

$$B = B_2 B_1$$
$$B_2 = U D$$ (4.58)
$$B_1 = V'$$

This can be interpreted as a feedforward linear net with one hidden layer

$$y = B_2 z$$
$$z = B_1 x$$

The interesting property of this structure is that the hidden layer activation vector is narrower than both the input and the output vectors, as depicted in Fig. 4.17. The hidden layer acts as a *bottleneck*, selecting H_{red} most informative combinations of input pattern combinations. So, this layer can be interpreted as a *feature extractor*.

This idea has its natural nonlinear counterpart in a nonlinear layered network. The degree of overdetermination can be determined following the principles explained in Sect. 4.2. The numbers to be compared are the number of constraints $K \times M$ and the number of free parameters P.

4.6   Statistical Conditions for Generalization

In addition to algebraic aspects of generalization, statistical properties of training and test sets are to be considered. These conditions concern the distribution of the training samples themselves, rather than the statistical properties of measurement noise discussed in the previous sections. From this viewpoint, the training samples are viewed as drawn from some real distribution. This distribution describes which patterns of the type relevant for the given application occur and with which relative importance. It is an empirical rather than probabilistic distribution and has not much to do with randomness: the only thing that may be viewed as random is which samples have been selected to be included in the training set, although also this can be done deliberately.

An example of this may be a set of images of human faces. The faces existing in the world and the ways to make their image are varying in some way that can be viewed as a distribution. However, a face is not genuinely random—what is random is which ones are included in an image collection used as a training set.

The science concerned with a deliberate selection of samples to receive optimal models is known as *design of experiments*. It has been developed primarily for the settings where the user has both the possibility and the necessity to construct experiments for gaining measurement data. This data is further used for model building. The discipline of design of experiments is concerned with making this selection in a way optimal from the application viewpoint—typically maximizing the accuracy of received models under given resource limitations. The focus is on the complete coverage of possible cases, without unintended bias toward untypical or irrelevant subsets.

There is plenty of sophisticated schemes for choosing the optimum experiments that might be considered for generating training sets. Wherever possible, appropriate schemes should be followed. This is the case if

- Measurement experiments can be planned.
- The dimensions of data patterns are small.
- The number of required data patterns (in face of conditions discussed in previous sections) is small.

Unfortunately, typical applications of contemporary DS do not offer many opportunities for applying this deliberate approach. One reason is in the data dimensions and volumes. Experimental schemes offered by the design of experiment theories are appropriate for pattern vectors of widths up to a few dozens. Beyond this size, a combinatorial explosion takes place that would lead to unrealistic training set sizes.

Another reason is the unknown characteristics of real-world distributions. It is hardly possible to characterize the distribution of pixels in a photograph of a human face or handwritten digit images. In particular, resorting to the well tractable Gaussian distributions can seldom be justified.

One more reason is the availability of data. A collection of handwritten digits (with assigned classes constituting the training set outputs) will not give the opportunity to be selected in a very deliberate way. It will typically contain pattern sets as they came in some data streams such as scans of postal envelopes.

All these reasons will frequently make it difficult to follow exact rational principles if collecting training sets. Ultimately, the only recommendation to be made is the following:

> The training set has to be representative of the whole data population in both
> scope and frequency distribution. This will have a decisive effect on the quality
> of the results obtained, in particular on the performance on the test set.

The requirement of being representative extends over all facets including time.
While linear models may allow a moderate amount of extrapolation, there is no
genuine extrapolation concept for nonlinear relationships of a priori unknown type.
This is why the pattern recognition materialized by the mapping extracted from
training data can only concern aspects in some way represented in that data.

A simple illustrative example is a model of fuel consumption of a car, gained from
the past data. It can make a forecast for fuel consumption in the following days if
the cruises take place in the same region with its typical cruise velocities, road turn
diameters and slopes. No forecast is possible if the model gained in a flat landscape
is applied to cruises across high mountains or in off-road conditions. The model can
also lose its validity after some years of car operation because of the wearing of car
components.

Similarly, the classification model of facial expressions with the help of images
of female humans probably cannot lead to good results for male humans.

Despite the trivial character of these recommendations, the aspect of good representations should be paid large attention—it will be crucial for the success of the
application.

4.7   Idea of Regularization and Its Limits

In DS literature (e.g., [3]), the concept of *regularization* is granted a lot of attention.
The motivation for regularization is similar to that discussed in Sects. 4.2–4.5. In the
underdetermined case with an insufficient number of training examples, the mapping
parameters determined with help of the training set are ambiguous—there is an
infinite space of possible solutions. In the presence of noise, undesirable fitting to
the noise takes place.

The availability of an infinite space of possible solutions suggests the possibility
of preferring some solutions to others, according to some a priori knowledge. The
idea of regularization is to use this knowledge for choosing the mapping parameters
without, or with less ambiguity.

The most explicit approach to regularization is with means of *Bayesian inference*. This is a method of parameter estimation. Alternative, non-Bayesian statistical
methods are, for example, the *maximum likelihood*, *least squares*, or the *moment*
method.

Bayesian inference is based on the Bayes formulas for joint and conditional probabilities. The joint probability of two phenomena A and B can be factorized as

$$P(A \cap B) = P(A \mid B) P(B) = P(B \mid A) P(A) \tag{4.59}$$

with $P(X)$ being the probability of phenomenon X and $P(X \mid Y)$ the probability of X conditional on occurrence of Y.

In parameter estimation, the question is how the parameter vector or matrix W can be determined from data D. In the setting of estimating a mapping between the input vectors organized in matrix X and output vectors in matrix Y, data D consists of the pair (X, Y).

Substituting D for A and W for B in (4.59) will deliver the equality

$$P(D \mid W) P(W) = P(W \mid D) P(D) \qquad (4.60)$$

Here, $P(D \mid W)$ is the conditional probability of the data instances in D given the parameters W. It is an alternative to the concept of likelihood of classical statistics. There, typical such parameters are, for example, means and variances of some distribution such as Gaussian. The conditional probability would then express the probability (or probability density) of data D randomly drawn from the density with parameters W.

$P(W)$ is the probability of certain parameters set according to the domain knowledge. It is called *prior probability*. In the Gaussian distribution, means with smaller absolute value and small variances may be most probable in some domains.

The goal of parameter estimation in the Bayesian sense is to determine the probability distribution $P(W \mid D)$ of parameters W given the observed data D. This probability is called *posterior probability*. From this probability distribution, the most probable configuration of W can be selected, for example, to implement the best mapping model. To do this, Eq. (4.60) is reformulated as

$$P(W \mid D) = \frac{P(D \mid W) P(W)}{P(D)} \qquad (4.61)$$

The probabilities in the numerator (the likelihood and the prior) are specified. The denominator can be received by norming over all $P(W \mid D)$ but is not required for determining the maximum probability for W.

Once the Bayesian optimum of parameters W is determined, the distribution

$$P(X \cap Y \mid W) \qquad (4.62)$$

of data $D = (X, Y)$ is specified. From this joint distribution of input X and output Y, the optimum mapping model is received from the conditional distribution

$$P(Y \mid X \cap W) = \frac{P(Y \cap X \mid W) P(W)}{P(X \cap W)} \propto \frac{P(Y \cap X \mid W)}{P(X \cap W)} \qquad (4.63)$$

This formula allows to determine the distribution of output Y for every given input X and optimal parameter set W. The mean of Y over this distribution is the Bayesian point estimate of the output.

The application of (4.61) and (4.63) is straightforward if all distributions involved are Gaussian. For many other distributions, no analytical expressions are known. At best, there are fitting pairs for likelihoods and priors such as binomial likelihood at beta prior. Without an analytical expression, the treatment of high-dimensional data, typical for the contemporary Data Science applications, is infeasible.

This unfortunate circumstance makes the well-founded Bayesian approach difficult to apply in practice. This has led to the development of simpler concepts where using domain knowledge is less explicit. The most popular of them is using an additional term in the objective function of model optimization. For example, the sum of square deviations can be extended by weighted regularization term:

$$\sum_k \left(f\left(x_k, w\right) - y_k \right)^2 + cr\left(w\right) \tag{4.64}$$

with a positive weight c. The regularization term is expected to have low values for more probable parameters w. The simplest and most widespread is a sum of squares of all model parameters (frequently neural network weights) w. This choice is justified by the assumption that small weights are more probable than large. Of course, this may apply in some circumstances and not so in others. A possible rationale behind this is that small weights lead to smoother mappings than large ones. Smoothness is a reasonable expectation for many application problems. Furthermore, smooth models are substantially easier to identify by numerical methods.

This simplified model has also a Bayesian explanation: the constant c is proportional to the prior variance of parameters if their prior distribution is assumed to be Gaussian with independent (i.e., uncorrelated) individual parameters.

More sophisticated regularization concepts consider the curvature (second derivative) of the parameterized mapping and try to minimize it. This concept is highly developed in the theory of *splines*.

Regularization addresses one of the most essential problems of learning. It is closely related to the concepts discussed in Sects. 4.1–4.6. In the underdetermined case, the parameter matrix based on input data matrix pseudoinverse is that with the smallest norm. This is related to the regularization concept of preference for small neural network weights. The difference between both is that the pseudoinverse is known explicitly and can be exactly computed, while using regularization, optimization with possible convergence problems takes place and the extent of regularization depends on the choice of regularization weight c.

It is important to point out that for underdetermined problems, the linear approach is usually that with the least number of parameters and thus with the least extent of underdetermination. Additionally, linear functional dependence is the smoothest one—its curvature is zero. So, in most applications, it is advisable to use linear models with pseudoinverse if the number of training examples is smaller than the input dimension. Alternatively, the approach omitting small singular values Sect. (4.5) can be used.

In overdetermined tasks, the noise considerations of Sects. 4.2 and 4.4 can support the need for regularization if the extend of overdetermination is low (i.e., a low ratio $\frac{K-N}{K}$) to alleviate the adverse effects of fitting to noise. It has to be kept in mind that is difficult to make general recommendations for the choice of both the regularization term and its weight, whose optimal value depends on the relative magnitude of unknown noise.

4.7.1 Special Case: Ridge Regression

A popular procedure with a good success record, frequently counted to the regularization methods, is the *Ridge Regression*, proposed by Hoerl and Kennard [5]. It is presented in this subsection as a representative for a group of similar methods such as Lasso regression (see e.g., [4]).

The original motivation for its development has been the treatment of strongly mutually correlated input variables, called *multicollinearity*. The regression formula (4.9) contains the an inverse of the matrix XX'. Multicollinearity leads to the term XX' being close to the loss of rank. Its inverse, determining the variance of the regression coefficients (4.48), is then close to infinity, that is, it contains large values. The numbers on the diagonal of the inverse determine the variance of the regression coefficients.

It has to be understood that the statistical view to regression frequently focuses on the accuracy of estimating the regression coefficients rather than on a small prediction error. The term *Mean Square Error* (MSE) is also frequently used for error of the coefficients, rather than the forecast error. This contrasts with the prevailing usage in DS, in particular neural network research, where MSE refers to the forecast.

To avoid the proximity to the singularity in the computation, the term XX' has been extended to

$$XX' + cI \qquad (4.65)$$

and the estimate of the regression coefficients to

$$B = YX'(XX' + cI)^{-1} \qquad (4.66)$$

instead of (4.9), with an appropriate positive constant parameter c. Setting $c = 0$, we receive the ordinary least square solution.

This amounts to the minimization of

$$\sum_k (bx_y - y_k)^2 + cbb' \qquad (4.67)$$

for each element of the output vector. This is a special case of the regularization of type (4.64).

A good survey of the properties of Ridge regression is given in [12]. The most interesting one is that the Ridge regression estimator of the "true" regression coefficients has (for some range of values of parameter c) a smaller MSE than the least square estimator with $c = 0$. This property is the result of a small variance of the Ridge regression estimator, outweighing the estimator bias (which is zero for least squares).

It has to be pointed out that this advantage of the Ridge regression estimator concerns only the expected MSE value over the statistical population. For the training set, the least square solution is always the optimum.

The range of parameter c for which this is true is between zero and the value of the term

$$\frac{2\sigma^2}{\beta\beta'} \qquad (4.68)$$

with σ being the standard deviation of the random error in the output data and β the true regression coefficient vector. This range grows with the extent of noise.

Knowing the range where the MSE of Ridge regression estimate is superior to that of least square estimate still does not quantify the extent of this superiority, that is, how large the improvement is. To assess this, let us observe the difference between both estimates. This difference is expressed by the matrix factor

$$W_c = (XX' + cI)^{-1} (XX')$$
(4.69)

Following the arguments around (4.51–4.53), this factor approaches the unit matrix for a constant c but growing number of training examples K. Consequently, the variance reduction is also vanishing with growing K. As the MSE of the Ridge regression estimator is additionally loaded by bias, the superiority of Ridge regression MSE diminishes with the growing size of the training set. In other words, Ridge regression provides essential advantage only for settings where the training set size is relatively close to the exactly determined case.

Ridge regression is also favorable in the case of high multicollinearity of input data. This is supported by the concept of degrees of freedom. They are intended to express the reduction of parameters that are effectively free by constraining their norm. They are defined in terms of the singular values d_i of matrix X:

$$\sum_i^N \frac{d_i^2}{d_i^2 + c}$$
(4.70)

In the overdetermined linear case, the rank of X is equal to the number of inputs N, and so is the number of singular values. Sometimes [1], this relationship is formulated in terms of eigenvalues of matrix $X'X$ which are equal to the squares of the singular values of X.

With $c = 0$, Eq. (4.70) is equal to N. For $c > 0$, the number of degrees of freedom is smaller than N. Then, it consists of summands that are particularly small if the corresponding singular value d_i is small. It is just the broad span of small and large singular values that constitutes the phenomenon of multicollinearity, leading to a large variance of parameter estimates. The number of degrees of freedom (4.70) might be used to assess the degree of overdetermination for a training set. The problem may be in determining the singular values for very large matrices X with a sufficient precision. Nevertheless, it provides insight into the mechanism of mean square reduction.

Ridge regression produces biased estimates. However, for a certain range of penalty parameter c, it can deliver a smaller MSE of parameter estimates over test sets than the least square procedure. This benefit grows with the amount of noise in the data and decreases with the number of training examples. So, the Ridge regression may be advantageous for problems that are only moderately overdetermined.

So far, the MSE of model parameter estimates has been discussed. However, in DS, particularly if treating huge data sets, the bias and variance of model parameters are of less interest. Neural network models with millions of parameters are scarcely viewed as mirrors of concrete "true" models down to the parameter level. Rather, a good fit and forecast capability is the key focus.

The mean square forecast error (in contrast to the mean square parameter error) suffers less from the phenomenon of multicollinearity. Let us image a trivial input/output relationship $y = x_1$, that is, a scalar output depending identically on the input x_1. Let us further assume a negligent user has included another input x_2 into the training set that is identical with x_1, $x_2 = x_1$. The matrix XX' consisting of parameter values of x_1 and x_2 is clearly singular and cannot be inverted. The MSE of least squares parameters (4.48) is infinite. Nevertheless, every model of the form

$$y = w_1x_1 + w_2x_2, \quad w_1 + w_2 = 1 \tag{4.71}$$

is a perfect forecast model—with zero forecast error. The individual regression coefficients w_1 and w_2 are not determined except for the condition $w_1 + w_2 = 1$. Every coefficient pair satisfying this condition is optimal in the sense of least squares of prediction error. In other words, in this case, the forecast error is not affected by the multicollinearity itself. Consequently, it is not guaranteed that the least square regression inferior to the Ridge regression in the mean square *parameter* error is also inferior in the mean square *forecast* error. Nevertheless, prediction superiority of Ridge regression is sometimes encountered for appropriate penalty parameter values.

It is difficult to determine the optimum penalty parameter analytically. Even the upper bound (4.68) of the interval in which its value is to be sought (for optimizing the mean square parameter error) cannot mostly be computed. Even less is known about the optimum for the mean square forecast error. This is why the only feasible procedures for this are based on *Cross-Validation*, discussed in Sect. 4.8. The simplest approach consists in

1. trying various values of c;
2. determining the optimum Ridge regression parameters for each value of c; and
3. selecting the optimal variant with help of a separate data set, the validation set.

4.8   Cross-Validation

In face of uncertainty how good forecasts for novel data not included in the training set may be, various schemes for assessment of the test set performance have been developed. They can all be summarized under the concept of *cross-validation*.

For least square procedures, which are free of meta-parameters, cross-validation delivers information how good the generalization is. The arguments of Sect. 4.4 suggest that the main uncertainty is for weakly overdetermined, or even underdetermined task settings. With strongly overdetermined tasks (with a number of constraints being a fair multiple of the number of free parameters), a convergence of training and test set mean forecast errors can be expected, as long as both sets are representative of the statistical universe of the application. In this sense, cross-validation can be seen as an useful option.

With regularization schemes such as Ridge regression, cross-validation is a necessary part of the training procedure. It is the only practical method how to fix the regularization parameters such as the penalty constants to values such that the regularization brings about more benefit than harm.

The simplest approach to cross-validation is to take out a part of the available training data (i.e., establishing a single test set, also called *validation set* in this context) and to use them after the accomplished training for evaluating the test set performance. This procedure assumes abundant data sufficient for both a representative training set and an equally representative test set.

Approaches more economical in data usage consist in making k data partitions and using them alternately as training and test data: k runs are accomplished in which one partition is the test set while the remaining $k - 1$ are used for training. This principle is shown in Fig. 4.18 with four partitions where individual runs correspond to the rows. Each training set consists of a relatively large fraction $\frac{k-1}{k} = 1 - \frac{1}{k}$ of the available data.

The extreme case of k equal to the number of training samples K is known as *leaving-one-out* scheme. It has the lowest bias and is related to the *Jackknife* procedure in statistics by Quenouille [10] and refined by Tukey [11]. With training sets of millions of samples, using this partitioning is clearly computationally infeasible since a complete optimization is to be executed in each run. Practical experience recommends between five and ten partitions. They can be sufficient for recognizing whether regularization (e.g., by Ridge regression) is helpful and with which penalty parameter values. For the latter decision, k runs are to be executed for every variant of the penalty parameter—which is computationally expensive enough.

Related, but more general, is the concept of *bootstrap* [2]. It consists in drawing L random subsets of size $k \leq K$ with replacement (i.e., the same individual sample can occur multiple times). The collection of L values of a statistical characteristic

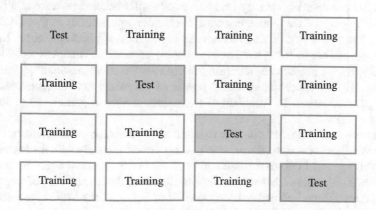

Fig. 4.18 Cross-validation with $k = 4$ data partitions

approximately mirrors the distribution of this characteristic in the statistical population. So, its statistical properties (mean, variance, etc.) can be assessed. However, general guarantees of the accuracy of such assessments by bootstrap are not available.

Unfortunately, characteristics such as MSE over individual training set samples suffer from the problem that bootstrap sets are overlapping and thus mutually dependent. This leads to an underestimation of the test set error. A discussion of this aspect is given in [4].

Bootstrap also suffers from the excessive computation burden. The recommended number of subsets is at least $L = \log (K) K$ [6]—mostly infeasible for large DS applications.

> Cross-validation concepts can help to assess the generalization performance of a model mapping. Furthermore, they are necessary for determining the appropriate variant of regularization parameters (discerning the values with which regularization is helpful from those where it is harmful). Cross-validation schemes delivering reliable statements (such as bootstrap) are computationally very expensive, up to infeasibility for large DS applications.

4.9 Parameter Reduction Versus Regularization

The arguments presented have disclosed the importance of an appropriate relationship between the number of free parameters of the mapping approximator (e.g., a neural network with its weights) and the training set size for the generalization performance. It can be said that an excessive number of parameters is the main reason for overfitting and thus for poor generalization.

On the other hand, as mentioned in concluding remarks of Sect. 4.3 and 4.4, an insufficient parametrization may lead to a poor approximation because of a lack of expressive power of the model. For complex model architectures as those based on neural networks, it is difficult to assess the sufficiency of parametrization. Mostly, it is only possible to manipulate some general parameters such as the widths of hidden layers. Sometimes, the decision principle is "the more the better", which may be productive for the representational power of the model but simultaneously bad for the generalization. The problem of excessive number of parameters is thus not avoided.

There are two basic methods how to avoid such excessive parameter sets. One of them is the direct one: to set the number parameters consistently with the recommendations of Sect. 4.1 in the linear case and Sect. 4.2 in the nonlinear one. Choosing the number of parameters such that it is substantially lower than the number of constraints prevents overfitting. Following the arguments of Sect. 4.4, the MSE approaches the inevitable noise level for both training and test sets.

The alternative approach has many supporters. It consists in taking an abundant number of parameters and addressing the risk of overfitting with the help of regularization, as discussed in Sect. 4.7. An example of typical argument is the following: " Generally speaking it is better to have too many hidden units than too few. With too few hidden units, the model might not have enough flexibility to capture the nonlinearities in the data ..." ([4], [11.5.4]). This is certainly true as long as no regularization has come into play. With regularization, this flexibility is restricted. It is the case even in the quantitative manner, taking into account the reduction of the number of degrees of freedom according to Eq. (4.70). Although it is not easy to capture the concrete consequences of this reduction, it is clear that the resulting flexibility in fitting the function to the data is reduced to some extent.

Lacking theoretical support, the behavior of both alternatives will be illustrated by a simple example, the approximation of a polynomial function as in Sect. 4.1. Let us have a fifth-order polynomial with six parameters. It can be exactly fitted to six training examples (points on the polynomial function). Our two alternatives are:

- Fitting the six training examples by a lower-order polynomial. To retain the character of the odd function, the third-order polynomial with four parameters has been selected.
- Fitting the same training examples by a higher-order odd polynomial and seeking the optimum regularization parameter c from Eq. (4.65). The optimum of c is sought by evaluating dense points on the given fifth-order polynomial function on the interval covered by the training set and determining the MSE. This is an analogy to taking an extensive verification set. For the optimum regularization parameter, the degrees of freedom according to Eq. (4.70) are determined.

The approximation by a non-regularized lower-order polynomial is plotted in Fig. 4.19. The MSE of this approximation over the test set is 0.034. The number of free parameters is four.

For regularized polynomials, two variants are depicted: regularized polynomials of seventh order (Fig. 4.20) and eleventh order (Fig. 4.21).

In both cases, an optimal value of the regularization parameter c has been determined, for which the verification set (referred to as "test set" in the further text and figure legends) MSE was minimal. The large values of this parameter result from large values of the higher powers of polynomial argument s. It can be observed that the optimum of the regularized seventh-order polynomial (MSE $= 0.039$) is close, although not as good as, than that of the non-regularized polynomial of third order. The degrees of freedom according to Eq. (4.70) are 3.45, which is even lower than those of non-regularized polynomial. Taking a regularized polynomial of the eleventh order (that is, with "abundant parameters") leads to clearly inferior results, with MSE $= 0.067$. In other words, even an optimized regularization does not help if the parameter number is excessive.

Of course, counterexamples can be found where regularized results are superior. A such case is depicted in Figs. 4.22, 4.23 and 4.24. Both the seventh- and the eleventh-order polynomials are, when regularized, superior in test set perfor-

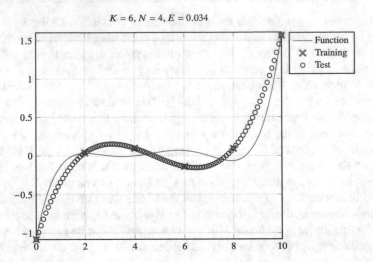

Fig. 4.19 Example 1: approximation by non-regularized polynomial of third order

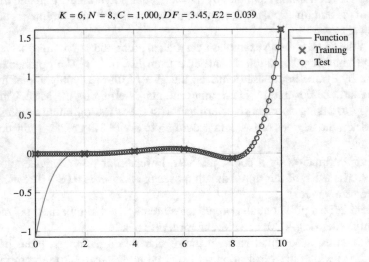

Fig. 4.20 Example 1: approximation by regularized polynomial of seventh order

mance to the simple non-regularized third-order polynomial. Nevertheless, the rich parametrization of the eleventh-order polynomial is not helpful.

Additionally, another unfavorable property of the more extensive parametrization can be observed in Figs. 4.25 and 4.26 presenting the dependence of the test set performance on the value of the regularization parameter c. While this dependence is monotonous for the seventh-order polynomial, it is not so for the eleventh-order one. Rather, there are two obvious local minima at about $c = 2e4$ and $c = 5e7$. This makes the search for the global minimum difficult: the only possibility is an enumeration of values of parameter c with a sufficiently small increment. More efficient search

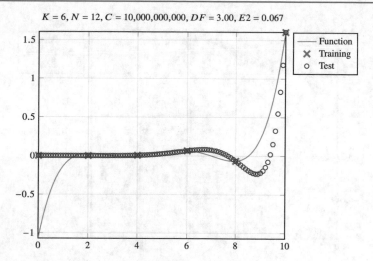

Fig. 4.21 Example 1: approximation by regularized polynomial of eleventh order

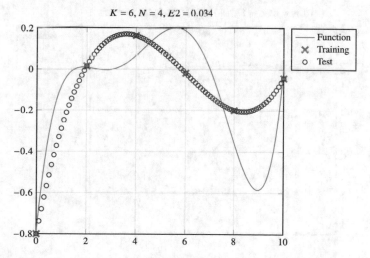

Fig. 4.22 Example 2: approximation by non-regularized polynomial of third order

methods such as bisection cannot be used without a risk of skipping over some of the local minima, maybe just the global one. Since, as stated in Sect. 4.7.1, this minimum has to be sought with the help of sufficiently large verification sets, it represents a formidable task.

Although these two examples provide no insight into the mechanisms of approximation in both approaches, they show the potential pitfalls. The approach characterized by

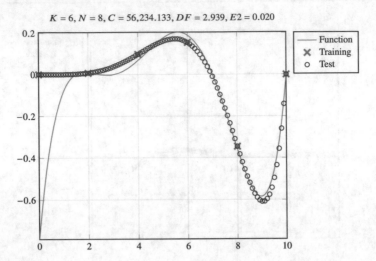

Fig. 4.23 Example 2: approximation by regularized polynomial of seventh order

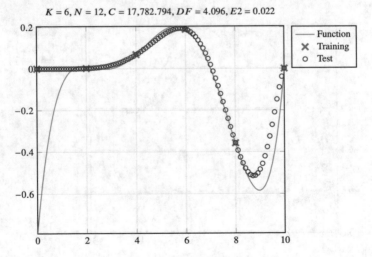

Fig. 4.24 Example 2: approximation by regularized polynomial of eleventh order

- providing a rich parametrization (which would correspond to the underdetermined case if there were no regularization); and
- regularizing the optimization task

may, or may not, be superior to the alternative approach with a limited parametrization that is explicitly overdetermined for the given training set size.

Nevertheless, there are some serious burdens connected with the regularization approach:

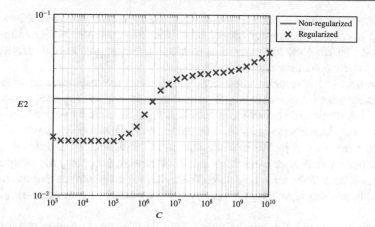

Fig. 4.25 Example 2: approximation by regularized polynomial of seventh order—test set MSE in dependence on the regularization parameter

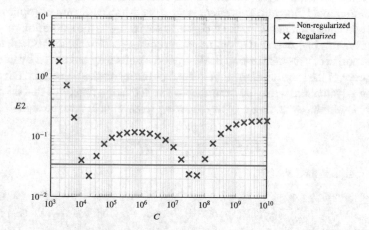

Fig. 4.26 Example 2: approximation by regularized polynomial of eleventh order—test set MSE in dependence on the regularization parameter

1. There are substantially more parameters to be optimized (e.g., 12 parameters for the regularized eleventh-order polynomial against 4 parameters for the non-regularized third-order polynomial.) This may be a key obstacle in contemporary Data Science tasks working with millions of parameters.
2. For the success of regularization, it is essential to find the optimal regularization parameter. Apart from the optimum, the test set performance may be arbitrarily poor. This optimum can only be found experimentally with the help of sufficiently large verification sets, recomputing the model for each value of the regularization parameter in a dense grid. This causes another computational burden that can be decisive for the success of the task.

Both factors together lead to a computationally very expensive method. Even with up-to-date computing equipment, it is difficult to guarantee MSE minimization solutions that ale close to the genuine minimum. But imprecise optimization may lead to the erosion of regularization benefits.

Sometimes, it is recommended (e.g., [1]) to start the minimization with small initial parameter values and to stop it prematurely as long as the parameters are still small. This would correspond to an "implicit regularization by poor convergence" since the regularization penalty term remains small. However, we have to keep in mind that the main objective is minimization of MSE, with an additional subgoal of parameter values remaining to be small. With early stopping, the subgoal may be reached but not the main objective. Then, there is no guarantee that the solution reached in this way is meaningful with regard to the MSE—the fit may be arbitrarily poor.

On the other hand, it has to be pointed out that the regularization penalty has a subsidiary effect, in particular, if used with neural networks. By constraining the weight values, it tends to prevent saturation of some nonlinear activation functions such as the sigmoid function. This effect is mostly desirable for numeric optimization reasons since the gradient w.r.t. the weights is vanishing in the saturation range.

However, this improvement of optimization convergence makes it difficult to evaluate the merit of the regularization itself—good results may be caused by improved convergence of the learning algorithm instead of the regularization effect. For a more conclusive evaluation, it may be desirable to use activation functions that do not saturate, for example, by adding a gently growing linear function to the sigmoid (3.3):

$$f(x) = \frac{2}{1 + e^{-x}} - 1 + ax \tag{4.72}$$

with a small constant a.

In summary, it is difficult to decide whether a modest, overdetermined parametrization or an abundant parametrization with regularization will lead to better generalization. From the viewpoint of computing resources, the former method is definitely superior.

4.10 Chapter Summary

Solving most DS problems consists in fitting a parameterized mapping to a set of measured input/output data pairs. Formally, a complete fit constitutes an equation system.

In this equation system, the number of variables is equal to the number of independent mapping parameters P. The number of equations equals to the number of training examples K multiplied with the number of output elements M.

For $P = KM$, the equation system is exactly determined (as long as parameters are mutually independent).

For $P < KM$, the system is overdetermined and has only an approximate solution.

For $P > KM$, the system is underdetermined and has an infinite number solutions, the choice between which is arbitrary.

It is meaningful to define the determination ratio $f = \frac{MK}{P}$. If the parameterized mapping is capable of approximating the reality from which the measured data have been drawn, some simple conjectures can be made. In the presence of noise, the following principles are valid if varying the size of the training set (while keeping the number of parameters unchanged):

- With the training data set used for parameter fitting, $f = 1$ allows a perfect fit to the noise, which is undesirable (since the noise is specific for the particular training set an not for the statistical population). With growing f, the fitting square error converges to the noise variance.
- With the test data (not used for parameter fitting), the expected square error is large for small f and converges, like for the training set, to the noise variance with growing f. Consequently, the model fit to the training and test set become close to each other with growing f. This is the substance of good generalization.
- With $f < 1$, no good generalization can be expected A straightforward remedy is to reduce the number of tree parameters to reach $f > 1$.

Varying the number of parameters (instead of the size of the training set), some variants may be insufficiently parameterized to capture the real mapping. Then, the square error minimum may, additionally to the principles listed above, grow with growing f.

These results are valid for unbiased estimates. Using biased estimates, it is, under certain conditions, possible to reduce the expected test set MSE in comparison with unbiased estimates. This approach is called regularization. The conditions for the potential test set error reduction are not easy to meet and require a repeated use of a representative verification set with the help of which the optimal regularization parameters are to be found. Otherwise, the test set performance may be substantially inferior to than for unbiased estimates.

4.11 Comprehension Check

1. Fitting a model to a set of K input/output pattern pairs (with input dimension N and output dimension M) can be viewed as an equation system. Of how many equations does it consist?
2. Assume a linear model or a linearization around a point in the parameter space. If the model contains P mutually independent parameters, what are the conditions for

     a. the exactly determined case with unique solution;
     b. the underdetermined case with an infinite space of solutions; and
     c. the overdetermined case without exact solution?

    Formulate the answers with the help of determination ratio f.

3. Under which conditions (in particular, concerning the ratio f) can the training and the test set square errors be expected to converge together?
4. Under which assumption would the training and test set square error converge to the variance of Gaussian noise?
5. Can the training set MSE with Ridge regression be expected to be below that of the unbiased least squares solution?
6. Is the potential advantage of Ridge regression larger with *strongly* or *weakly* correlated input feature?
7. Which practical procedure can be used for determining optimal regularization constant such that the biased estimate with Ridge regression is better than the unbiased one?

References

1. Bishop CM (2006) Pattern recognition and machine learning. Information Science and Statistics, Springer, New York
2. Efron B (1979) Bootstrap methods: another look at the jackknife. Ann Stat 7(1):1–26. https://doi.org/10.1214/aos/1176344552
3. Goodfellow I, Bengio Y, Courville A (2016) Deep learning. Adaptive Computation and Machine Learning. The MIT Press, Cambridge
4. Hastie T, Tibshirani R, Friedman J (2009) The elements of statistical learning. Springer series in statistics. Springer, New York. https://doi.org/10.1007/978-0-387-84858-7
5. Hoerl AE, Kennard RW (1970) Ridge regression: biased estimation for nonorthogonal problems. Technometrics 12(1):55–67. https://doi.org/10.1080/00401706.1970.10488634
6. Kendall MG, Stuart A, Ord JK, Arnold SF, O'Hagan A (1994) Kendall's advanced theory of statistics, 6th edn. Edward Arnold, Halsted Press, London, New York
7. Kohavi R, Wolpert D (1996) Bias plus variance decomposition for zero-one loss functions. ICML
8. Moore EH (1920) On the reciprocal of the general algebraic matrix. Bull Am Math Soc 26:394–395
9. Penrose R (1955) A generalized inverse for matrices. Math Proc Camb Philos Soc Camb Univ Press 51:406–413
10. Quenouille MH (1949) Approximate tests of correlation in time-series 3. Math Proc Camb Phil Soc 45(3):483–484. https://doi.org/10.1017/S0305004100025123
11. Tukey J (1958) Bias and confidence in not quite large samples. Ann Math Statist 29:614
12. van Wieringen WN (2021) Lecture notes on ridge regression. http://arxiv.org/abs/1509.09169

Numerical Algorithms for Data Science

5

In Chap. 2, standard application-specific mappings were shown. The task of DS is to find the mapping exhibiting the best correspondence to a set of training data. To identify the best correspondence or fit, appropriate quantitative characteristics have been formulated. Mostly, they have the form of an error measure that is to be minimized. For some of them, a perfect fit is characterized by the error measure being zero, while others (in particular those for classification tasks) do not have this advantageous property.

Once an error measure is selected, the search after the best mapping consists in finding mapping parameters minimizing the error measure over the training set. For some formulations with linear mappings (or, more exactly, mappings linear in parameters), an explicit analytical form of the optimal parametrization is known. Prominent examples are the least squares problem minimizing the MSE over the training set and the log-likelihood minimization, both mentioned in Chap. 2. The algorithms for these cases get along with standard operations of linear algebra. For their solution, efficient and reliable algorithms are known and widespread in various mathematical, engineering, statistical, and Data Science tools. They are scaling to sizes of millions of parameters.

For mappings nonlinear in parameters, explicit solutions are scarcely available. Then, minimizing the error measure has to be performed by a numerical search. Neural networks presented in Chap. 3 count to this category. As they became a significant Data Science tool, numerical minimization algorithms are playing a key role in DS.

The necessity of finding a solution very close to the genuine minimum goes beyond the obvious desire for error being "the less, the better". Some theoretical properties are given only if the error minimum is exactly reached. These theoretical properties are mostly strictly proven only for linear systems, but analogies can be expected for nonlinear ones, too. The minimum of MSE, frequently sought in problem classes of Chap. 2, guarantees the estimate to be unbiased. This is not the case apart from the minimum. Advantageous properties of biased estimates, e.g., those

© The Author(s), under exclusive license to Springer Nature Switzerland AG 2023
T. Hrycej et al., *Mathematical Foundations of Data Science*, Texts in Computer Science,
https://doi.org/10.1007/978-3-031-19074-2_5

provided by Ridge regression of Sect. 4.7.1, are reached only in the minimum of the regularized error measure. Classification follows Bayesian principles (Sect. 2.2) only in the minimum of corresponding terms. These are additional reasons why numerical optimization methods should exhibit good convergence to the minimum. They also should not be terminated prematurely; otherwise, the properties of the solution may be arbitrarily far away from the expected ones. This is why these algorithms are presented in some detail in this chapter.

5.1 Classes of Minimization Problems

Minimization problems can be partitioned into several classes defined by the characteristics of the function to be minimized and additional requirements on the solution. Knowing the class of the problem is essential for choosing the appropriate algorithm capable of solving it.

One criterion for classification of minimization problems is by the presence or absence of additional constraints to the optimized parameters. These basic classes are

- unconstrained minimization and
- constrained minimization.

Constraints usually have the form of equalities or inequalities between some algebraic terms containing the parameters. In DS, the only prominent case of using constraints is the SVM mentioned in Sect. 2.2.1. The minimization problem formulation consists of constraints (2.29) and minimizing the term (2.30).

The quadratic norm (2.30) makes this task a quadratic minimization problem with linear constraints. Its solutions are the subject of the mathematical discipline called *quadratic programming*. There are several efficient solving algorithms implemented in software packages. A popular method with polynomial time scaling (i.e., solving time polynomially growing with the number of parameters) is the *interior-point method* [4]. The applied variants are surveyed by Potra and Wright [18]. Mathematical foundations of quadratic programming are rather complex. Their relevance only for a small, although important, subset of Data Science problem formulations renders them beyond the scope of this chapter.

Apart from this problem formulation including constraints, most other DS problems are defined as unconstrained minimization problems. This includes the popular algorithms used with neural networks. This is why an overview of some important principles of unconstrained numerical minimization is given in the following subsections. They are ordered by generality, starting with an important special case of quadratic optimization that provides the basis for solution of more general problem classes. To be precise, only optimization problems with continuous parameters (i.e., parameters that can acquire continuous values) are considered here. The voluminous domain of integer or discrete optimization is omitted since it has no particular relevance for typical Data Science problems.

5.1.1   Quadratic Optimization

The structurally simplest function with a finite minimum is the quadratic function, or a polynomial of order two. Polynomials of lower order than two are the constant function (order zero) and the linear function (order one). While the former is trivial, the latter has no extreme, that is, neither minimum nor maximum as it is growing and falling to infinity.

A multivariate quadratic polynomial has the general form

$$f(x) = \frac{1}{2}x'Ax - bx \tag{5.1}$$

If matrix A is non-singular and positive definite, the function (5.1) has a unique minimum. This minimum is where the gradient with respect to vector x is zero:

$$\frac{\partial f(x)}{\partial x} = \nabla f = Ax - b = 0 \tag{5.2}$$

resulting in

$$Ax = b \tag{5.3}$$

or

$$x = A^{-1}b \tag{5.4}$$

This form of the gradient term justifies the widespread use of the factor $\frac{1}{2}$ and a negative sign of the linear term.

In addition to the possibility of using the explicit expression (5.4), the minimum of (5.1) can also be found by iterative minimization. The advantage of this approach is that matrix A and vector b need not be explicitly known. Instead, the gradient (5.2) is to be computed. Although the gradient can be approximated numerically as a vector with scalar elements

$$\frac{\partial f}{\partial x_i} \approx \frac{f(x_i + \Delta x_i) - f(x_i - \Delta x_i)}{2\Delta x_i} \tag{5.5}$$

with appropriate small finite differences Δx_i, this procedure is particularly beneficial if the gradient is known as an analytical expression.

Interestingly, the gradient-based procedure may be equally fast, and sometimes more accurate, than computing the matrix inversion in (5.4). The algorithms of this type are discussed in Sect. 5.3.

5.1.2   Convex Optimization

The class of optimization problems next to the quadratic optimization is *convex* optimization. This discipline of numerical mathematics provides algorithms that can efficiently solve minimization problems with a function $f(x)$ convex in argument vector x. A scalar function of this type is such that all function points are above its tangents, as shown in Fig. 5.1. Vector functions are such that function points lie above all tangent hyperplanes.

Fig. 5.1 Scalar convex
function

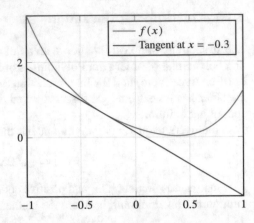
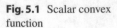

Fig. 5.2 Contour lines of a
2D convex function

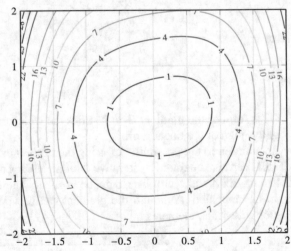

In two dimensions, it is the characteristic that the level curve density decreases
with decreasing function values, as shown in Fig. 5.2.

For the class of convex functions, the gradient-based algorithms mentioned in
Sect. 5.1.1 can be used very efficiently. They scale polynomially with the problem
size. Convex optimization can be viewed as the best understood discipline in numerical optimization.

5.1.3   Non-convex Local Optimization

The next in difficulty is the class of *non-convex* functions with a single local minimum. Mentioning the last property is crucial since this specifies the difference to the
most difficult class referred to in Sect. 5.1.4. A local minimum is defined as a point
such that in its (maybe small) environment all function values are larger than at the
minimum point.

Fig. 5.3 Non-convex scalar function with a single local minimum

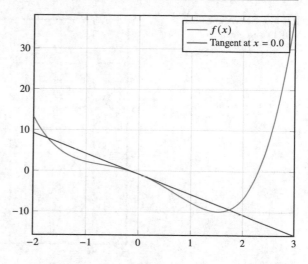

Fig. 5.4 Non-convex 2D function with a single local minimum

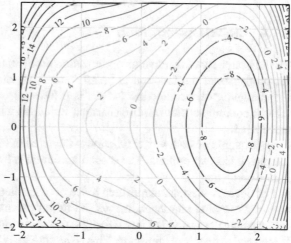

A scalar example is shown in Fig. 5.3. It is the characteristic that there are points in which the tangent crosses the function plot, lying above it in some segments. There is only one local minimum, positioned at $x = 1.5$.

In two dimensions, the level curve density may increase with decreasing function value at some regions (see Fig. 5.4).

Non-convexities reduce the efficiency of algorithms for convex optimization. Nevertheless, most of these algorithms still converge to the local minimum. They exploit the property of this function class that for every starting point, there is a continuous monotonically descending path to the minimum.

Fig. 5.5 Scalar functions
with two minima

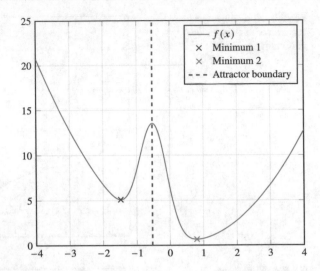

5.1.4　Global Optimization

The last category of continuous optimization problems makes no assumptions about the properties of the function to be minimized, although most approaches assume the continuity of the function (which is different from assuming that parameters take on continuous values). In particular, functions with multiple local minima are considered.

An example of such scalar function is depicted in Fig. 5.5. In the range shown in the plot, it has two minima. The second minimum is lower and is thus referred to as *global minimum*.

For obvious reasons, algorithms based on a gradient descent would converge to one of these minima, depending on the initial parameter value. The range of the scalar parameter is divided into two regions, referred to as *attractors*, each of which is assigned to one local minimum. This is the case for all gradient descent algorithms, no matter whether their convergence is fast or slow.

For multidimensional parameters, the problem is the same, with potentially complex boundaries reminding of class boundaries discussed in Sect 2.2. An example in the two-dimensional parameter space is shown in Fig. 5.6.

It is clear that multiple minima are possible only if the function is not convex. This is a feature common with the class discussed in Sect. 5.1.3. There, local optimization of a possibly non-convex function is the solution approach. This local minimum is also a global one, whenever there is a single attractor. This is not the case for functions addressed in the present section which are characterized by the presence of multiple attractors.

The 2D example suggests also the additional dimensionality problem. For a scalar function, the *density* of minima depends on the ratio of the absolute values of the first and second derivative. This ratio expresses the relative rate of direction change. The larger this rate, the more local minima can arise on a given interval of parameter

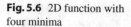

Fig. 5.6 2D function with four minima

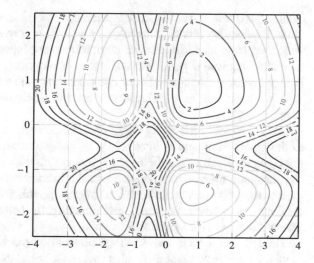

values. This is the case for every dimension, so the number of potential minima with the same bound of the derivative ratio will exponentially grow with the number of parameters.

The lack of advantageous scaling with the number of parameters has led to a plenty of approaches, some of which attempt to provide some kind of convergence guarantee. However, this is only possible for a relatively small number of parameters not beyond about ten.

A large part of these approaches are working with a simultaneously processed set of solution "individuals", each of which may potentially be pursued with the help of convex optimization algorithms. To minimize the number of convex optimizations, individual solutions are clustered to groups such that the group members probably count to the same attractor. Then, it is sufficient to apply convex optimization to a single, appropriately chosen individual. This is the substance of clustering methods [20]. Other approaches attempt to provide probabilistic guarantees of convergence: Rinnooy Kan and Timmer [20] or Mockus et al. [15].

Also the recently popular genetic algorithms are working with sets (or "populations") of solutions [9]. This group of algorithms seems to be relatively easy in handling but scarcely provides convergence guarantees for problems of a size usual in Data Science.

The fact of exponential growth of the number of potential local solutions with problem dimension makes clear that the population-based approaches will always lag behind the real problem complexity—exponentially growing solution populations are not tractable for Data Science problems whose size frequently reaches the order of magnitude of millions.

There is a persistent discussion about the seriousness of the problem of local minima in the neural network community. On one hand, it is clear that even small networks with nonlinear units may have multiple minima. On the other hand, it is known that there is a substantial redundancy of solutions. A layered neural network such as that of Fig. 3.4 has equivalent parametrization for every permutation of hidden units: Exchanging hidden units together with their corresponding input and output weights retains identical network output for every input. For H hidden units, the number of equivalent parameter sets is thus $H!$ for every hidden layer. For multiple hidden layers, the total number of equivalent parameter sets is the product of $H_i!$. This is a huge number with layers of sizes around thousand, not unusual in recent applications.

To summarize, performing global optimization with a guarantee of convergence to the global minimum is infeasible with today's means. On the other hand, it is unknown to which extent the local minima are inferior to the global one. In result, the pragmatic solution is to perform a local minimization (which is computationally expensive enough for large neural networks), possibly from several initial points. As there are no guarantees for global optimality, the solution is accepted or rejected after validation in the application domain.

This is the reason why only local optimization approaches will be further discussed in this chapter. Extensions alleviating (but not solving) the multiple minima problem are addressed in Sect. 5.5.

5.2 Gradient Computation in Neural Networks

For all classes of optimization problems, the majority of algorithms makes use of the gradient of the objective function. The gradient is the vector of derivatives of the objective function (e.g., squares of deviations between the measured and forecast output pattern value) with regard to individual parameters of the functional approximator (e.g., weights of a neural network). Although the gradient can be also determined numerically by computing the objective function values for small changes of each parameter as in (5.5), it is substantially more efficient to make use of an analytical form of the gradient. In the special case of neural networks, there is a particular technique for computing the gradient called *backpropagation*. Backpropagation is a straightforward application of the chaining rule for derivatives, but its importance for the widely used layered neural networks justifies mentioning it explicitly. The name is due to the fact that the gradient is computed backward from the objective function over the network output toward the network input.

The point of origin of the computation is the vector of derivatives of the objective function w.r.t. network output $\frac{\partial E}{\partial y}$. As a convention throughout this book, it is a column vector. For the MSE objective function, it would be a vector with elements $2\left(y_i - y_{ref,i}\right)$.

The h-th layer of a layered network can be formally described as a vector function

$$z_h = F_h \left(W_h z_{h-1} + b_h\right) \tag{5.6}$$

with $F_h()$ being the vector activation function of the h-th layer consisting of identical scalar activation functions $f_h()$ applied to individual argument elements one by one.

For elements z_{hi} of layer output vector z_h, (5.6) corresponds to

$$z_{hi} = f_h \left(\sum_{j=1}^{n_h} w_{hij} z_{h-1,j} + b_{hi} \right) \tag{5.7}$$

In the output layer, z_h is identical with the output vector y.

To determine the objective function value, the expressions (5.6) are to be chained along growing h and the result inserted as argument to the objective function. This is frequently referred to as *forward pass*.

The gradient of objective function w.r.t. parameters of this layer is

$$\begin{aligned}
\frac{\partial E}{\partial w_{hij}} &= \frac{\partial E}{\partial z_{hi}} \frac{\partial z_{hi}}{\partial \hat{z}_{hi}} z_{h-1,j} = f' \frac{\partial E}{\partial z_{hi}} z_{h-1,j} \\
\frac{\partial E}{\partial b_{hi}} &= \frac{\partial E}{\partial z_{hi}} \frac{\partial z_{hi}}{\partial \hat{z}_{hi}} = f' \frac{\partial E}{\partial z_{hi}}
\end{aligned} \tag{5.8}$$

with $\hat{z}_{hi}$ being the argument of activation function $z_{hi} = f(\hat{z}_{hi})$.

As a matrix expression, it is

$$\begin{aligned}
\frac{\partial E}{\partial W_h} &= \Delta F_h \frac{\partial E}{\partial z_h} z'_{h-1} \\
\frac{\partial E}{\partial b_h} &= \Delta F_h \frac{\partial E}{\partial z_h}
\end{aligned} \tag{5.9}$$

where ΔF_h is a diagonal matrix with diagonal elements equal to the derivatives of the scalar activation functions of the h-th layer. The gradient of objective function w.r.t. input vector z_{h-1} to this layer with elements $z_{h-1,j}$ is, analogically

$$\frac{\partial E}{\partial z_{h-1,j}} = \sum_{i=1}^{m_h} \frac{\partial E}{\partial z_{hi}} \frac{\partial z_{hi}}{\partial \hat{z}_{hi}} w_{hij} = \sum_{i=1}^{m_h} f' \frac{\partial E}{\partial z_{hi}} w_{hij} \tag{5.10}$$

or, as a matrix expression using column vectors z

$$\frac{\partial E}{\partial z_{h-1}} = W'_h \Delta F_h \frac{\partial E}{\partial z_h} \tag{5.11}$$

Chaining the expressions (5.9) for all successor layers (along decreasing layer index h) and substituting into the expression (5.11) indexed by h represents the gradient w.r.t. h-th layer. This is called *backward pass*. This arrangement is efficient: The computational expense for the backward pass is only double of that for the forward pass, once by the application of Eq. (5.9) and once by Eq. (5.11). To determine the gradient of objective function, both a forward and a backward pass are to be performed. By contrast, to determine the functional value, the forward pass is sufficient.

5.3 Algorithms for Convex Optimization

The basis for convex optimization algorithms is the minimization of a multivariate quadratic function (5.1). Its minimum is at the point where the gradient (5.2) is zero so that the condition (5.3) is satisfied. Let us suppose matrix A and vector b are known.

The simplest approach is to determine the gradient by Eq. (5.2) and descend in the direction opposite to the gradient (since the gradient gives the direction of the maximum growth of the function) by some step length. The best step to do is to go, in the descent direction, to the point where the function (5.1) has the lowest value. This can be done by a procedure called *line-search*. It is a one-dimensional minimization in a given direction. There are several efficient algorithms for line-search with well-defined convergence and accuracy.

The behavior of a simple gradient descent algorithm with a simple two-dimensional quadratic function with minimum function value zero is observed in Fig. 5.7.

The descent path intuitively suggests an unnecessary large number of steps. In fact, after five search steps, the function value (listed in Table 5.1) is still 0.00119, while the minimum is exactly zero.

The reason for this is the following: After having found the minimum in the search direction u, the gradient v is orthogonal to u (since there is nothing to be improved in direction u), that is

$$u'v = 0 \tag{5.12}$$

However, proceeding in direction v, the original gradient will change by

$$Av \tag{5.13}$$

and so partially spoil the progress in direction u. This results in the requirement that the gradient change has to retain this progress. This is attained by the condition of the change being *conjugate* (a generalization of the orthogonality property) to u, or

Fig. 5.7 Steepest descent of a 2D quadratic function

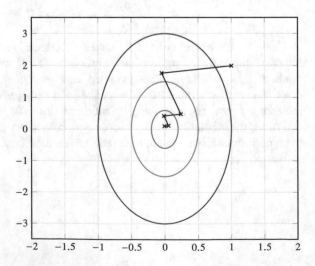

Table 5.1 Function values attained by steepest descent

Iteration	Function value
1	0.34925
2	0.08444
3	0.02042
4	0.00494
5	0.00119

Table 5.2 Function values attained by descent in conjugate directions

Iteration	Function value
1	0.349
2	0.000

$$u'Av = 0 \tag{5.14}$$

Consequently, the gradient g_k received in the k-th iteration is corrected to the vector h_h conjugate to previous descent directions. This is algebraically done by the correction subtracting the conjugate projection of g_k to the space spanned by previous search directions:

$$h_k = g_k - \sum_{i=1}^{k-1} a_i h_i = g_k - \sum_{i=1}^{k-1} \frac{h_i' A g_k}{h_i' A h_i} h_i \tag{5.15}$$

This excludes the gradient component that is contained in the space of these previous search directions.

With a known matrix A, it is not necessary to determine the optimum step length with the help of line-search. The iteration in terms of parameter vector x is

$$x_{k+1} = x_k - \lambda_k h_k = x_k - \frac{h_k' g_k}{h_k' A h_k} h_k \tag{5.16}$$

with λ_k being the analytically known optimum step length.

Proceeding in such corrected gradient directions will result in the path presented in Fig. 5.8.

This procedure attains the exact minimum in only two steps (Table 5.2).

The knowledge of matrix A and vector b would be a requirement difficult to satisfy in most applications including the minimization of an unknown convex function. To account for this, the practically used version of conjugate gradient method assumes no knowledge of matrix A and vector b. Then, the constant λ_k in (5.16) is no longer analytically known. Instead, it determined with the help of line-search—it is the step length with which the minimum in the search direction is attained. Interestingly, also the conjugate projection (5.15) results directly from the optimal λ_k, leading to the

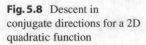

Fig. 5.8 Descent in conjugate directions for a 2D quadratic function

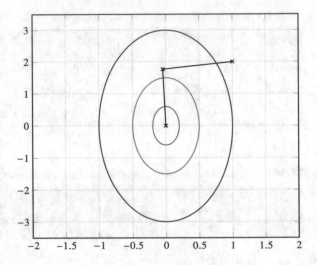

algorithm of Fletcher and Reeves [6] with a substantial improvement by Polak and Ribière [17]. Both versions and their implementations are explained in [19].

With non-quadratic but convex functions, matrix A corresponds to the Hessian matrix (i.e., the matrix of second derivatives)

$$A = \left[a_{ij} \right] = \left[\frac{\partial^2 f}{\partial x_i \partial x_j} \right] \tag{5.17}$$

The outstanding property of the second-order methods in general and of the conjugate gradient method in particular is their *quadratic convergence* in the environment of the minimum: the number of significant digits to which the solution is determined doubles every iteration. The formal definition of this property is that

$$\| f_{k+1} - f_{\min} \| \approx C \| f_{k-1} f_{\min} \|^2 \tag{5.18}$$

This relationship concerns convex but non-quadratic functions. Remind that for exact quadratic functions of dimension N, N iterations are sufficient for the exact solution (up to machine precision).

Whether quadratic convergence is reached also far away from the minimum depends on the similarity of the function to a quadratic one. In other words, if the residue in the Taylor expansion of the function up to the second-order term is small, the convergence is close to quadratic. In large real-world problems, it is more realistic to expect a linear convergence

$$\| f_{k+1} - f_{\min} \| \approx C \| f_{k-1} f_{\min} \| \tag{5.19}$$

In spite of the non-spectacular term "linear", this is good enough. It implies that the distance between the attained and true minimum decreases as a geometric progression, that is, exponentially.

Another beneficiary property of the conjugate gradient is its low memory requirement. It typically needs only three vectors of the length equal to the number of parameters. This is important for large DS applications, in particular those using neural networks with millions of network weights.

Conjugate gradient algorithms and corresponding line-search methods are implemented in many numerical packages.

There are some other approaches to numeric convex optimization with quadratic convergence. A popular class is *quasi-Newton* methods, also called *variable metric* methods [19]. They are based on the principle of successively approximating the matrix inverse A^{-1} needed to determine the minimum via (5.4). Their convergence reliability is similar to that of the conjugate gradient. However, their memory requirement is quadratic with the number of parameters N since the complete matrix inverse of size $N \times N$ has to be stored.

Another approach is specific for minimization in least squares problems, which are widespread in DS. The error function with parameter vector denoted, for consistence with previous explanations, as x, is

$$f(x) = \sum_{k=1}^{K} \left(y_k - g(u_k, x) \right)^2 \tag{5.20}$$

Its Hessian matrix (5.17) consists of elements

$$
\begin{aligned}
a_{ij} &= \frac{\partial^2 f}{\partial x_i \partial x_j} \\
&= \sum_{k=1}^{K} \frac{\partial^2 \left(y_k - g(u_k, x) \right)^2}{\partial x_i \partial x_j} \\
&= -2 \sum_{k=1}^{K} \frac{\partial \left(\left(y_k - g(u_k, x) \right) \frac{\partial g(u_k, x)}{\partial x_i} \right)}{\partial x_j} \\
&= 2 \sum_{k=1}^{K} \left(\frac{\partial g(u_k, x)}{\partial x_i} \frac{\partial g(u_k, x)}{\partial x_j} - \left(y_k - g(u_k, x) \right) \frac{\partial^2 g(u_k, x)}{\partial x_i \partial x_j} \right) \\
&= 2 \sum_{k=1}^{K} \frac{\partial g(u_k, x)}{\partial x_i} \frac{\partial g(u_k, x)}{\partial x_j} - 2 \sum_{k=1}^{K} \left(y_k - g(u_k, x) \right) \frac{\partial^2 g(u_k, x)}{\partial x_i \partial x_j}
\end{aligned}
\tag{5.21}
$$

It can be argued that the second term approaches zero in the proximity of the minimum since there, the fitting errors $y_k - g(u_k, x)$ will be averaged out. Then, the Hessian matrix can be approximated by the single term

$$a_{ij} = 2 \sum_{k=1}^{K} \frac{\partial g(u_k, x)}{\partial x_i} \frac{\partial g(u_k, x)}{\partial x_j} \tag{5.22}$$

for which only the first derivatives are sufficient. It can be directly used for determining the approximate parameter minimum (5.4) of the present iteration. This is the substance of the *Levenberg–Marquardt* method [12,13]. This method requires memory of size $N \times N$, too.

5.4   Non-convex Problems with a Single Attractor

DS mapping tasks using a mapping nonlinear in parameters are not guaranteed to be convex minimization problems. This is why the convergence properties of second-order optimization methods may substantially degrade.

The non-convexity can have different forms. For large data sets and complex mapping approximators such as deep neural networks, round-off errors may lead to local "roughness" of the function so that numerically computed gradients make the function appear non-convex. But even smaller problems may lead to function of types depicted in Figs. 5.3 and 5.4. They may (but need not) exhibit non-convexities in some points while remaining to be convex in others. Non-convex regions tend to be remote from the minimum, while the environment of the minimum is convex. It has to be emphasized that the existence of a convex region around the minimum is certain, by definition of minimum of a smooth function. This is also obvious from the square error properties, as supported by the relationships (5.20), (5.21), and (5.22). By contrast, the existence of non-convex regions is just a possibility, depending on the concrete form of the nonlinearities of the mapping to be approximated with the help of the square error criterion.

Local non-convexities, if encountered, are a problem if the Hessian matrix (the matrix of second derivatives) is explicitly or implicitly used in the algorithm. In non-convex regions, it is not positive definite. A possible remedy is to enhance the Hessian matrix A by adding a weighted unit matrix:

$$A_{\mathrm{corr}} = cI + A \qquad (5.23)$$

The corrected matrix A_{corr} is then used in computations. For the conjugate gradient algorithm which does not use explicit Hessian matrix, there are simple algorithmic modifications preventing bad convergence in such points. However, quadratic convergence is not guaranteed in non-convex regions.

Anyway, algorithms for convex optimization of Sect. 5.3 are still good candidates for non-convex problems with a single local minimum. On the other hand, the huge advantage of the second-order methods over the first-order ones is reduced or may completely disappear at such possibly occurring non-convex regions. This has led to a widespread turning away from the second-order methods in the DS community. Whether this is justified is a difficult question that cannot be answered at the present time. It is only possible to compile some arguments.

Those in favor of the first-order methods are the following:

- As stated above, the convergence speed advantage of the second-order methods is reduced by the existence of non-convex regions in the parameter space.
- First-order methods are simpler to understand and implement, although the difference to the second-order methods is not large.
- First-order methods are less sensitive to numerical imprecisions.
- The benefit of using line-search to find the accurate minimum in the first-order gradient direction is much smaller than with second-order methods, for which it is sometimes a part of the algorithmic idea. This offers the possibility to use a

simpler and computationally less expensive concept of fixed or adaptive gradient steps after every gradient evaluation.

- The benefit of using exact gradient is also smaller with first-order methods. Frequently, it is then computed only approximately with the help of a small subset of the training set. This also leads to saving computing time.

On the other hand, the following arguments may support doubts whether the listed arguments are conclusive:

- In by far the most application cases, the attainable minimum of the error measure is not known. Consequently, the merits of individual first- or second-order algorithms cannot be objectively compared—it is difficult to say whether the minimum obtained is "sufficiently good".
- First-order methods show only moderate benefits from machine precision. This has led to a frequent use of hardware with single-precision floating-point arithmetic such as graphic cards. But then, no objective comparison with second-order methods which have higher requirements on computing precision is possible. This may lead to a kind of "vicious circle": Commitment to hardware inappropriate for the second-order methods prevents their appropriate use.
- Poor convergence of first-order methods sometimes leads to oversized neural networks. Increasing the network size can help the convergence if the method is converging as badly as being close to random—it is then equivalent to using a pool of solutions "located" in various subnetworks (e.g., parts of oversized hidden layers). Some of these "regional" solutions may occur to be good fits, while other subnetwork regions are obsolete for the fit. For example, in a hidden layer of length 10,000, units with index 1001–1200 happen to materialize a good fit to data, while those with indices 1–1000 and 1201–10,000 are obsolete by delivering insignificant activation values for training examples. With a powerful optimization algorithm, 200 hidden units would be sufficient. Using an excessive number of free parameters may lead (or is almost sure to lead) to poor generalization (see Sect. 4.5).
- Huge size of some prominent applications such as BERT [3] or *VGG19* [23] makes it difficult to compare different methods because of required computing resources. The method used by the authors is the frequently assumed to be the optimal one, without further proof or discussion.

The resolution of this methodical dichotomy will not be easy because of a limited number of studies. The investigation of Bermeitinger et al. [2] using problems with a known minimum disclosed superiority of the conjugate gradient over several popular first-order methods for medium size networks, but with an advantage gap diminishing with size. So, it cannot be reliably extrapolated to expectations about very large networks.

First-order methods lack a unifying theory—the theory of numeric optimization unambiguously favors second-order methods. The merits of the first-order methods are mostly experimental. This is why only the basic principles of popular approaches will be introduced in the following subsections, without clear recommendations for their choice.

5.4.1 Methods with Adaptive Step Size

Once the gradient of the error function at given parameter vector values is known, this knowledge is maximally exploited by line-search which delivers the minimum in the direction of descending gradient. For conjugate gradient algorithms, this applies to the corresponding conjugate gradient direction. Under the commitment to use simple steepest gradient descent, the benefit of line-search is less pronounced—it does not prevent the algorithm from following the zigzag path as in Fig. 5.7. Then, it is worth considering to refrain from line-search at all. The alternative is to proceed in the descending direction (opposite to the gradient w.r.t. the parameter vector w) by a fixed amount c:

$$w_{t+1} = w_t - c\nabla E_t \tag{5.24}$$

or, for individual vector elements

$$w_{t+1,i} = w_{t,i} - c\frac{\partial E\left(w_{t,i}\right)}{\partial w_{t,i}} \tag{5.25}$$

(The term ∇E_t refers to the gradient of error function E at the t-th iteration.)

Such a fixed step length has been used in some pioneering algorithms such as *Adaline* (developed by Widrow and Hoff [24]).

For a simple quadratic function $E = aw^2$ with the derivative $\frac{dE}{dw} = 2aw_0$ at point w_0, the optimal step is w_0 so that the optimal step length parameter c satisfies the equality

$$c\frac{dE}{dw} = 2caw_0 = w_0 \tag{5.26}$$

resulting in a constant optimal parameter

$$c = \frac{1}{2a} \tag{5.27}$$

Unfortunately, this applies only to normalized quadratic functions, a very scarce subspecies of all convex functions. Even then, the constant optimal value exists but is not known a priori, depending from the unknown parameter a. So, although using a constant step size is appealing, it clearly cannot be optimal for all problems and all phases of the optimization.

The principle of following the steepest gradient is also scarcely the best way, as explained in the discussion of the conjugate gradient method (see Sect. 5.3). This has motivated modifications of first-order gradient descent where both the step length and the direction are modified. A frequently followed principle is making relatively longer steps in directions of parameters with a small derivative.

AdaGrad [5] tries to enforce larger steps for parameters with a small derivative through weighting. Weights are equal to the reciprocal value of the derivative vector norm, concatenated over past iterations.

$$w_{t+1,i} = w_{t,i} - \frac{c}{\sqrt{d_{t,i}}} \frac{\partial E\left(w_{t,i}\right)}{\partial w_{t,i}}$$

$$d_{t,i} = \sum_{s=1}^{t} \left(\frac{\partial E\left(w_{t,i}\right)}{\partial w_{t,i}} \right)^{2}_{t,i} \tag{5.28}$$

The larger the norm $\sqrt{d_{t,i}}$, the smaller the factor scaling the step in the i-th direction. **RMSprop** [8] uses a similar principle, taking an exponential moving average (with decay) of the derivative square

$$w_{t+1,i} = w_{t,i} - \frac{c}{\sqrt{d_{t,i}}} \frac{\partial E\left(w_{t,i}\right)}{\partial w_{t,i}}$$

$$d_{t,i} = g d_{t-1,i} + (1-g) \left(\frac{\partial E\left(w_{t-1,i}\right)}{\partial w_{t-1,i}} \right)^{2} \tag{5.29}$$

Adam [10] uses a weight consisting of the quotient of the exponential moving average derivative and the exponential moving average of the square of the derivative

$$w_{t+1,i} = w_{t,i} - \frac{cm_{t,i}}{\sqrt{d_{t,i}}} \frac{\partial E\left(w_{t,i}\right)}{\partial w_{t,i}}$$

$$m_{t_i} = g_1 d_{t-1,i} + (1-g_1) \frac{\partial E\left(w_{t-1,i}\right)}{\partial w_{t-1,i}}$$

$$d_{t,i} = g_2 d_{t-1,i} + (1-g_2) \left(\frac{\partial E\left(w_{t-1,i}\right)}{\partial w_{t-1,i}} \right)^{2} \tag{5.30}$$

All these algorithms have their supporters and adversaries. For example, Adam has gained popularity through being used in the developer team of the successful concept of the Transformer architecture (see Sect. 6.7).

5.4.2   Stochastic Gradient Methods

Classical statistical modeling is usually designed to deliver optimal fit for a given sample set, which is equivalent to the term training set used in the neural network community.

Early approaches counting to the neural network or AI domain such as the perceptron of Rosenblatt [22] have gone another way. They have been supplied training patterns one by one, updating the weight parameters after each pattern.

This idea was justified by the ambition of being *adaptive*. For an assessment of potential of such approaches, it is important to briefly discuss the substance of adaptiveness.

A system is usually considered as adaptive if it is able to adjust itself to changing environment or conditions. An adaptive image classifier designed to discern images of chairs from those of tables would have to be able to work if originally wooden chairs are successively substituted by metal ones, or by chairs without back rest. An adaptive cruise controller of a compact car would be expected, after some gradual retraining, to be able to control a heavy truck.

Although adaptiveness defined in this way is certainly a useful property, it is not easy to reach. It is substantially more difficult than to design a classifier or a controller for a stationary population of patterns fixed in advance. Also, their theoretical properties (such as the convergence of the perceptron for separable classes mentioned in Sect. 2.2.1) are mostly proven only for the stationary case. This is why such potentially adaptive concepts are usually used under stationary conditions.

Under such conditions, the term "adaptive" is somewhat misleading. Instead, the term *incremental* is more appropriate. An incremental algorithm makes parameter updates after each sample pattern. Such algorithms have also been developed in classical statistical modeling [1]. Some of them are constructed for their results to be equivalent to their "batch" counterparts. For example, parameters for an input/output mapping optimal in the sense of MSE are given by Eq. (4.5), i.e., $B = YX^{-1}$. *Recursive Least Squares* consist in keeping record of successive updates of both Y and X, performing matrix inversion after each sample. With the help so-called matrix inversion lemma, the computing expense for the inversion can be reduced. Such incremental algorithms always provide a current optimal solution in real time. On the other hand, in other than real-time settings, they provide no particular advantage and produce additional overhead of repeated matrix inversion updates.

This computing overhead can be avoided at the expense of sacrificing equivalence to batch algorithms. Supposing training samples are drawn from a stationary population generated by a fixed (unknown) model, the *stochastic approximation* principle can be used. It has been discovered by Robbins and Monro [21] in the context of finding the root (i.e., the function argument for which the function is zero) of a function $g(x)$ that cannot be directly observed. What can be observed are randomly fluctuating values $h(x)$ whose mean value is equal to the value of the unobservable function, that is,

$$E\big[h(x)\big] = g(x) \tag{5.31}$$

Fitting an input/output mapping to data by gradient descent is such a task. For mapping parameter vector w, the gradient $h(w)$ with respect to the error function computed for a single training sample has an expected value equal to the gradient $g(w)$ over the whole data population. The local minimum of the error function is where the $g(w)$ and thus the mean value of $h(w)$ are zero.

Robbins and Monro [21] have proven that the root is found with probability one (but without a concrete upper bound of number of necessary updates) if the parameter is updated by the rule

$$w_{t+1} = w_t - c_t h(w_t) \tag{5.32}$$

Fig. 5.9 Separation example for batch versus incremental learning

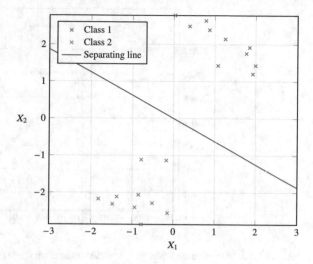

and if the step size sequence c_t satisfies the following conditions:

$$\sum_{t=1}^{\infty} c_t = \infty$$

$$\sum_{t=1}^{\infty} c_t^2 < \infty$$

(5.33)

A sequence satisfying conditions (5.33) is, for example, $c_t = \frac{1}{t}$. The former condition is necessary for the step not to vanish prematurely before reaching the optimum with a sufficient precision. The second condition provides for decreasing step size. With a constant step size, the solution would infinitely fluctuate around the optimum. In the context of error minimization, the reason for this is that although the gradient $g\,(w) = E\,[h\,(x)]$ will be gradually vanishing as approaching the minimum, its random instance $h(x)$ will not diminish for individual samples. The obvious cause for this is that gradients of individual samples do not vanish even if their mean over the training set is zero. At the minimum, $g(w) = 0$ will be the result of the balance between individual nonzero vectors $h(w)$ pointing to various directions.

To illustrate the behavior of incremental learning, let us consider a simple example of separating two classes, as depicted in Fig. 5.9.

These classes, consisting each of ten samples, can be separated by minimizing the MSE (2.26), the deviation between the linear class forecast and the true class indicators $(-1, 1)$. The classes are symmetric around the origin, so the choice of symmetric class indicators allows to omit the absolute term (*bias parameter*). The separating line shown in Fig. 5.9 corresponds the minimum MSE $E_{min} = 0.72484$, with weight vector $w_{min} = \begin{bmatrix} w_1 & w_2 \end{bmatrix} = \begin{bmatrix} 0.22167 & 0.35605 \end{bmatrix}$.

Gradient descent in batch mode (i.e., computing the gradient over the whole training set of 20 samples) follows the weight path shown in Fig. 5.10. The values

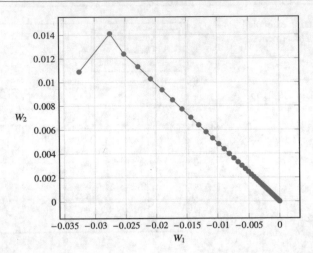

Fig. 5.10 Classification example: weight evolution in batch mode (weights relative to the optimum)

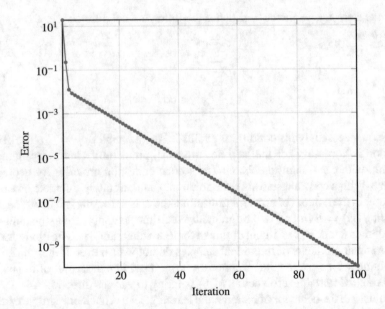

Fig. 5.11 Classification example: error evolution in batch mode

of both weights are depicted relative to the optimum weight, i.e., $w - w_{min}$, so that the optimum is reached at the vector $[0\ 0]$.

The step size has been set to the best value that leads to no oscillations: $c = 0.005$. The evolution of the MSE relative to its minimum E_{min} is shown in Fig. 5.11. A precision of 10^{-6} is reached after about 50 iterations.

Fig. 5.12 Classification example: weight evolution in incremental mode (weights relative to the optimum), constant step size

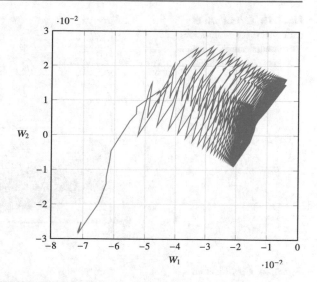

In the incremental mode with 2000 incremental iterations (which correspond, with a total of 20 samples, to 100 batch iterations), the weights do not converge to a stable state (Fig. 5.12). Such convergence also cannot be expected since the incremental gradient will never converge to a zero norm. The relative error is fluctuating, too. After 2000 incremental iterations, it is still considerable, fluctuating between 0.002 and 0.03. Considering only the errors after fully accomplished batches (i.e., every 20 increments), it reaches the minimum relative value of $E - E_{min} = 0.00275$. The error development is shown in Fig. 5.13.

Better results are attained with a step size schedule consistent with (5.33). With decreasing step size, the fluctuations are decreasing in amplitude (Fig. 5.14). However, the relative error (Fig. 5.15) is even larger than that with constant step size after full batches: $E - E_{min} = 0.00819$.

Using SGD is widespread in the present neural network community. The arguments in its favor are mostly focusing on the low computing effort for a parameter update based on a single training example. Goodfellow et al. [7] argue that the standard deviations of the gradient components are decreasing with the square root $\sqrt{K}$ of number of samples used, while the computing expense is increasing with K. This is, of course, true for independent samples drawn from a population. However, in a usual learning setting, these samples are drawn from the training set (which itself is a sample) and not from the whole population. The gradient computed over the whole training set in the batch mode (i.e., whole epochs) is the exact true gradient for the given training set, while the incremental gradients are only its approximations.

The additional argument cited is that what is genuinely sought is the minimum for the population and not for the training set. This argument is correct but does not really support the use of stochastic gradient algorithms. There is no method for finding the true, exact minimum for the population only on the basis of a subsample such as the training set—the training set is the best and only information available. Also, the

Fig. 5.13 Classification
example: error evolution in
incremental mode, constant
step size

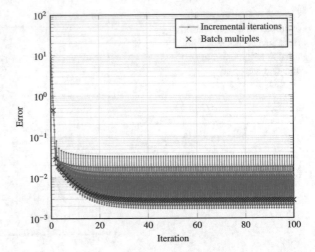

Fig. 5.14 Classification
example: weight evolution in
incremental mode (weights
relative to the optimum),
decreasing step size

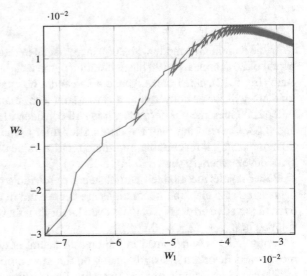

error function values used in the algorithm are values for the given training set. The
example of this subsection shows that there is no law guaranteeing computing time
savings through incremental learning for the same precision of the result.

Nevertheless, SGD may be economical in settings where the training set is so large
that its subset is sufficient for identifying the model. However, then, the possibility
of reducing the training set to a representative subsample and batch training with this
subsample should be considered. This can be done in the first training phase to find a
good initial solution. This phase can be followed by fine-tuning using the complete
training set.

It can be shown that first-order algorithms are doing better with SGD than the
second-order ones. This results from the fact that the Hessian matrix is more sensi-

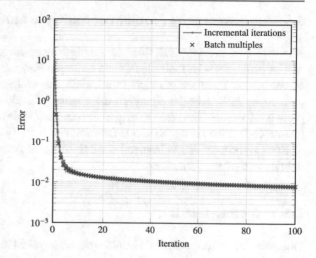

Fig. 5.15 Classification example: error evolution in incremental mode, decreasing step size

tive to random errors than the gradient itself. This will strongly reduce the performance of the second-order algorithms which use implicitly or explicitly the Hessian matrix. The influence on the first-order algorithms is weaker since they consider only the gradient. This is sometimes used as an argument against the use of second-order algorithms. Also, this conclusion is not strictly justified—the only conclusive consequence is that the second-order algorithms should not be used together with stochastic gradient. All three remaining alternatives

• the first-order algorithms with stochastic gradient;
• the first-order algorithms with batch (i.e., epoch based) gradient; and
• the second-order algorithms with batch gradient

remain unaffected.

SGD methods converge if the conditions of properly diminishing step size are satisfied. The argument in favor of their use is reducing computing expense to determine the gradient. This gain can in many cases be more than lost by a convergence inherently slower than for batch algorithms. Additionally, SGD is not compatible with the use of second-order optimization methods. In problems with very large training sets for which appropriate subsets can be expected to be sufficiently representative, such subsets can be directly used for a complete training to save computations.

5.5   Addressing the Problem of Multiple Minima

As mentioned in Sect. 5.1.4, there are no algorithms for genuinely global optimization of problems with multiple minima having a size usual in real-world problems. Nevertheless, there are attempts to alleviate the multiple minima problem. Their philosophy is to give the algorithm the chance of leaving one attractor and to explore another one. The phrase "to give the chance" is deliberately used—there is no guarantee that the change to another attractor will really take place.

Such algorithms are widespread and exhibit good performance record, in spite of the absence of any guarantees.

5.5.1   Momentum Term

One group of such algorithms is inspired by the physical idea of a mass body rolling or gliding over a rough surface. Without *momentum*, this body would stop at the bottom of any, even the smallest, groove or dent in the surface. However, the mass momentum makes the body cross small grooves and finally stop at the bottom of a groove that is deep enough that the inertial energy of the body is not sufficient to climb up the groove side. (From the physical point of view, the presence of friction continually diminishing the inertial energy would be a necessary condition. Otherwise, the body would cross all symmetric grooves of any size.)

The implementation of this idea does not genuinely follow the physical prototype. It consists in using a modified gradient h_t in the t-th step instead of the actually computed gradient g_t. The modification is based on a *memory* of past gradients:

$$h_t = (1 - c) \, h_t + c g_t \tag{5.34}$$

The positive constant c is $c \ll 1$, so that the actual gradient g_t has only partial influence on the modified gradient h_t.

To be aware of the properties of this appealing construct, it is useful to point out its relationship to a low pass filter known in signal processing [16]. Using the notation usual in signal processing, the low pass filter is typically used for filtering away the high-frequency noise of signal $u(x)$, delivering the filtered signal $y(x)$. The discrete variant of the simplest form of the low pass filter is formally identical to (5.34):

$$y(x + \Delta x) = (1 - c) \, y(x) + c u(x), \quad 0 < c < 1 \tag{5.35}$$

The argument x is frequently time although spatial filters where x corresponds to the position are also widespread. The increment Δx corresponds to the discrete processing step (the reciprocal value of the signal sampling rate).

In momentum terms, x can be viewed as the position in the parameters space over which the objective function is minimized, $u(x)$ is the gradient at x, and $y(x)$ the smoothed gradient. In contrast to the signal filter, which is usually updated at equidistant time points, the gradient is evaluated at successive optimization steps, which do not take place at points equidistant with regard to the position x. However, the operation principle remains the same.

Fig. 5.16 Bode plot of a low
pass filter

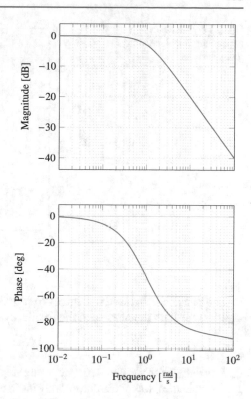

For the following analysis, the objective function surface can be seen as a com-position of sinusoidal components with various frequencies. The rough form of the function corresponds to some low frequency, while noise grooves have high fre-quencies. The low pass filter in (5.35) has the property of reducing the amplitude of higher frequencies contained in the signal while leaving low frequencies unchanged. This property is consistent with the goal of the momentum term: following the rough form of the objective function while ignoring the noisy grooves. The price for the amplitude reduction is the delayed phase of the high-frequency signal. As will be seen in the following text, this is an unfavorable property.

Both the amplitude reduction and phase delay for a sinusoidal input in depen-dence on the frequency are shown in Fig. 5.16, known as *Bode plot* in the control community [25]. The cutoff frequency of the filter, where the amplitude reduction starts, is arbitrarily chosen as 1 Hz. The qualitative behavior of the filter is analogical for other cutoff frequencies.

The upper part of Fig. 5.16 shows the amplitude on the logarithmic scale (20 dB correspond to a factor of ten). The amplitude remains untouched up to approximately the cutoff frequency. For higher frequencies, it becomes substantially reduced. The phase shown in the lower part is close to zero (i.e., no delay) for low frequencies but grows to negative values (i.e., some delay) for higher frequencies.

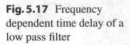
Fig. 5.17 Frequency
dependent time delay of a
low pass filter

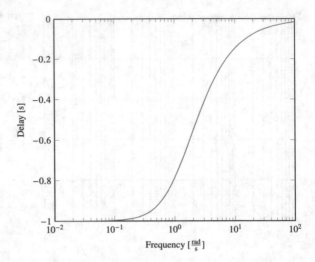

The phase is expressed in degrees (out of 360). Using the same units (degrees or radians) for both the phase and the frequency, the delay in time (or space) units is

$$\frac{\omega_{\text{ph}}}{\omega} \tag{5.36}$$

The delays for various frequencies are depicted in Fig. 5.17. We can see that for low frequencies, where the phase delay is close to zero, the absolute delay in seconds is the largest. For the frequency of $1\,\frac{\text{rad}}{\text{s}} = 57.296\,\frac{\circ}{\text{s}}$ (which corresponds to $\frac{1}{2\pi}\,\text{Hz} = 0.15915\,\text{Hz}$), the input signal amplitude is reduced by only 3 dB, that is, by factor 0.707. But the phase delay is already substantial: $45°$. This corresponds, for the frequency $1\,\frac{\text{rad}}{\text{s}}$, still to a considerable delay of $\frac{\pi}{4}\,\text{s} = 0.79\,\text{s}$.

These relationships are the same if the low pass filter is applied to space units (such as position) instead of time. In the case of momentum, the counterpart of time variable is the locations in the directions of the gradient, that is, the direction of descent steps. The delay corresponds to a shift in the position against that where the gradient has been determined. The example of Fig. 5.17 shows the following. The spatial shift of our smoothed function (corresponding to the absolute time delay above) is still large even for frequencies that are only moderately reduced (and thus are not smoothed away). So, the smoothed function is considerably distorted.

The idea of the momentum term is to provide a "smoothed" gradient without fluctuations caused by high-frequency disturbances. For simplicity, let us observe a function to be minimized that consists of a scalar curve. It is composed of a decreasing straight line and a sine disturbance:

$$u(x) = -ax + b\sin(\omega x) \tag{5.37}$$

The derivative of this function is

$$\frac{\mathrm{d}u(x)}{\mathrm{d}x} = -a + \omega b\cos(\omega x) \tag{5.38}$$

Fig. 5.18 Linear descent function with sinusoidal noise

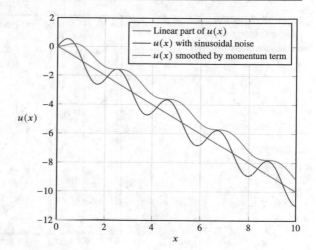

With $\omega b > a$, the negative derivative of the linear descending function becomes occasionally positive and would lead to a wrong direction of the gradient descent.

With a momentum filter, we receive

$$\frac{dy(x)}{dx} = -a + \omega br(\omega)\cos\left(\omega x + \omega_{ph}(\omega)\right) \qquad (5.39)$$

with the specific amplitude reduction factor r and phase ω_{ph} both depend on frequency ω.

Now, the condition for the occurrence of the locally growing derivative is

$$\omega br(\omega) > a \qquad (5.40)$$

which can be expected to occur considerably scarcer if the reduction factor r is small (i.e., if there is a substantial reduction of amplitude). In the example given in Fig. 5.18, no growing derivative of the smoothed function occurs although it does for the non-smoothed function.

In the upper panel of the Bode plot in Fig. 5.16 (the amplitude plot), the x-axis (the frequency) is logarithmic. The y-axis uses the decibel unit (dB), which is a logarithm of amplitude transfer r (20 dB correspond to a factor of 10). In the concrete plot of Fig. 5.16, this dependence is obviously linear for frequencies over about 2 rad/s. In this frequency range, the dependence of the logarithm of amplitude on the logarithm of frequency can be written as $\log(r) = h - d\log(\omega)$, or $r = \frac{e^h}{\omega^d}$.

So, it is obvious from the Bode plot (and theoretically proven for this type of low pass filter) that for high frequencies the reduction factor is inversely proportional to the frequency so that $\omega^d r(\omega)$ is a constant. For the first-order low pass filter of (5.35), the power d is equal to one and the product $\omega r(\omega)$ is a constant. Then, beyond a certain frequency (depending on slope a and noise amplitude b), the condition (5.40) is violated and no growing derivative is encountered. Consequently, avoiding the growing derivative (which would lead to stopping the gradient descent) depends only on the slope a and the disturbance amplitude b. In other words, on a steep basis function slope, even relatively deep noise grooves will still leave the smoothed function descending.

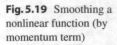

Fig. 5.19 Smoothing a nonlinear function (by momentum term)

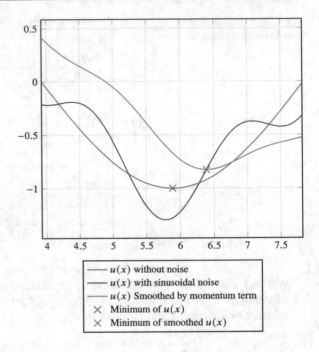

——	$u(x)$ without noise
——	$u(x)$ with sinusoidal noise
——	$u(x)$ Smoothed by momentum term
✕	Minimum of $u(x)$
✕	Minimum of smoothed $u(x)$

To summarize, if

- the downward slope a of the function to be minimized (without noise);
- the frequency ω; and
- the amplitude b of the noise

were known, it would be possible to determine the momentum parameter c in (5.34) such that the descent algorithm would not get stuck an local minima caused by the noise.

Of course, in reality, the mentioned constants are varying over the minimized function and can hardly be credibly assessed. At first glance, a possible strategy would be to take a very small value of c, filtering out a very broad frequency spectrum of noise. Unfortunately, this would cause a serious problem connected with distorting the minimized function.

As stated above, the momentum filter causes a spatial shift of the function, an analogy to the delay of a signal filter. In other words, the smoothed derivative y refers to a different point x than the actual one. Within a linear segment of the underlying objective function, the effect of this spatial shift is not observed since the linear function has an identical derivative everywhere. However, this is not the case for a nonlinear function with a local minimum, as illustrated by Figs. 5.19 and 5.20.

Figure 5.19 shows a nonlinear convex function with sinusoidal noise, Fig. 5.20 its derivative. The harmful effect of the spatial shift due to the phase caused by the momentum term can be observed. The sign change of the derivative is at a different

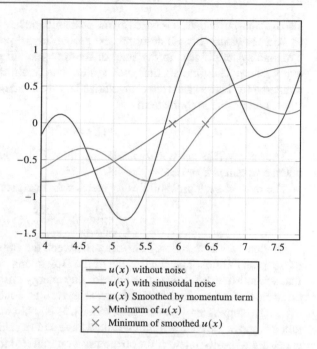

Fig. 5.20 Derivatives of the smoothed nonlinear function (by momentum term)

place (farther right) for the smoothed derivative than it is the case for the original derivative. Consequently, the minimum of the smoothed function is displaced. This is certainly a problem for the descent algorithms. The extent of this problem is influenced by the frequency spectrum of the function fluctuations considered to be "noise" and the local steepness of the "essential" error function to be minimized. It is difficult to asses these effects quantitatively and to assign appropriate values to the descent algorithm hyperparameters. Nevertheless, the momentum term has been successfully used in many applications. Anyway, it is important to be aware of its properties and possibly make experiments with various momentum parameters.

> The momentum term operates in a way analogous to a low pass filter. It is used in neural network training algorithms for smoothing the gradient of a non-convex function. The low pass analogy discloses problems connected with this mode of use—in particular, the phase delay leads to determining the gradient "out of place". This concept works only if the curvatures of the "essential objective function" and of the "disturbances" can be well assessed.

5.5.2 Simulated Annealing

There is one more approach addressing the problem of multiple local minima with roots in physics. In contrast to the concept of momentum that is easily imaginable also for non-physicists, it is based on the probabilistic law of thermodynamics known as *Boltzmann statistics*. It describes the probability of possibles states of a thermodynamic system if it is in the state of thermodynamic equilibrium under a given temperature. This implies that such system can attain diverse states for the same temperature. These states are characterized by their energy. The probability of each state is proportional to the term

$$P(E) \approx e^{\frac{-E}{kT}} \tag{5.41}$$

The energy E is given in *Joule*, temperature T in *degrees Kelvin*, and the Boltzmann constant is $k = 1.380649e{-}23$.

The ratio of such probabilities for states with energies E_0 and E_1 is

$$\frac{P(E_1)}{P(E_0)} = e^{\frac{-E_1 - E_0}{kT}} \tag{5.42}$$

Obviously, it is the energy difference between the states that determines the ratio of the probabilities. The small value of the Boltzmann constant makes it also clear that this ratio can be close to one only for tiny energy difference: Even for the energy difference of one billionth of a Joule, it is zero at double machine precision (for everyday temperatures on earth). In other words, only energy states at a very small amount above the energy minimum state have a noticeable probability. Such energy states are typically relevant for phenomena on molecular or atom particle level.

However, the property of the Boltzmann distribution important for optimization is the dependence of the probability ratio on the temperature. This principle does not depend on scaling, so constants and scales can be chosen to be more illustrative. For a better tractable constant $k = 1$, the probability ratios at (artificially defined) temperatures of $T = 1$ and $T = 10$ are depicted in Fig. 5.21. States with a slightly higher-energy level are quite probable at "temperature" $T = 1$, but their probability vanishes at energy difference of about seven. By contrast, at the higher temperature of $T = 10$, the distribution is more balanced—even a change to the state with energy larger by a factor of ten has still a considerable probability.

There is a physical process using this principle for a kind of optimization. This process is called *annealing* and consists in slow cooling of a material, starting at a relatively high temperature. The high-temperature phase makes various states (e.g., crystalline organization of the material) accessible even if the path to them crosses higher-energy levels. In this way, a rough global optimization is performed. The low-temperature phase makes the material to settle in the lowest-energy state that can be found in the neighborhood of the result of the high-energy phase. The slow pace of cooling is a precondition for the success of this process.

The annealing principle has attracted the attention of scientists working on hard problems of global optimization. The pioneering algorithm has been developed by Metropolis et al. [14]. A more general formulation has been proposed by Kirkpatrick et al. [11] and applied to *NP*-complete traveling salesman problem of discrete optimization (search for the shortest path to visit a set of locations).

The computing implementation of the annealing idea is not completely straightforward. While the state probabilities are fixed and the random state selection is

Fig. 5.21 Probability ratios at different temperatures

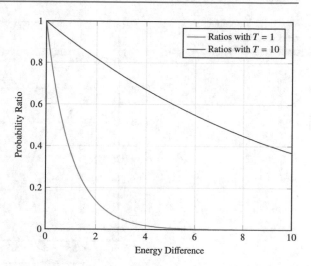

being done by natural laws in genuine annealing, both have to be materialized in some defined way in the computing algorithm. The typical way is to select the next solution randomly. This solution is accepted with certainty if it is better than the current one. If it is not, it is accepted only with a certain probability, depending on the extend to which is is worse (in the sense of the objective or error function) than the current one. The probability schedule is subject to "cooling": Accepting worse solutions becomes less probable during the algorithm progress, which corresponds to lowering the imaginary temperature.

In the corresponding chapter of Press et al. [19], it is argued that such procedure is not very efficient since it generates many worse solutions (that are probably rejected) among a very few better ones. They propose to combine the probabilistic step selection with a simple numeric optimization algorithm.

There are some variants of this idea. The gradient search can be combined by perturbing the solution by a small amount if the algorithm does not move. Beyond the intuition that one can have "good luck" to cross the boundary between local attractors, such perturbations can be viewed as smoothing operators. In the simplest case of a uniform distribution of the random perturbation Δx on an interval, that is, the probability density of Δx being

$$p\,(\Delta x) = \frac{1}{2d}, \quad \Delta x \in \langle -d, d \rangle \tag{5.43}$$

the mean value of such randomly perturbed error function value corresponds to the gliding average of width $2d$. For an appropriate value of d, the effect is similar to that of the momentum term. The function of Fig. 5.19 with sinusoidal noise is then smoothed as in Fig. 5.22, its derivative being that of Fig. 5.23.

The striking property of such smoothing is that it is more symmetric than with the momentum term. The minimum of the smoothed error function is much closer to that of the non-smoothed one. For some types of noise such as white noise, they are even identical.

Fig. 5.22 Smoothing a nonlinear function by uniform distribution

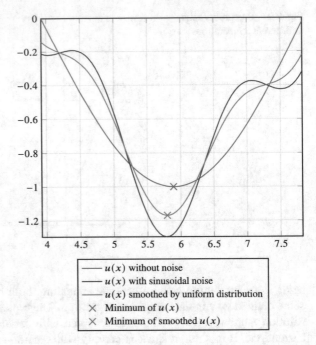

- $u(x)$ without noise
- $u(x)$ with sinusoidal noise
- $u(x)$ smoothed by uniform distribution
- ✕ Minimum of $u(x)$
- ✕ Minimum of smoothed $u(x)$

Fig. 5.23 Derivatives of a nonlinear function smoothed by uniform distribution

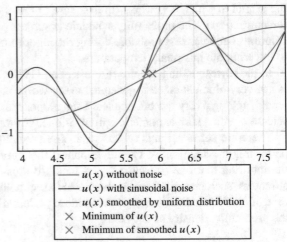

- $u(x)$ without noise
- $u(x)$ with sinusoidal noise
- $u(x)$ smoothed by uniform distribution
- ✕ Minimum of $u(x)$
- ✕ Minimum of smoothed $u(x)$

On the other hand, the computing expense for this is substantially higher. It is necessary to evaluate the error function at random points around the current solution. For high dimensions of the parameter space (as usual in DS applications), the expense would be excessive. This is a point in favor of the momentum term.

An interesting, practical, and popular possibility has already been mentioned in another context in Sect. 5.4.2. SGD is doing exactly this. The gradient over the whole training set is equal to the expected value of the gradients of individual training samples. Computing the gradient for individual training set samples and using it for parameter updates is similar to random perturbations. These perturbations correspond to deviations of sample gradients from the expected value. The analogy to the concept of simulated annealing goes even further. The condition of diminishing gradient step (5.33) requires that these random perturbations become gradually smaller. This, in turn, is also the principle of simulated annealing embedded in the schedule of decreasing "temperature".

This makes the expectations of efficiency of simulated annealing similar to those of SGD, with all their advantages and deficiencies. The success depends on the choice of annealing schedule and other process parameters.

5.6 Section Summary

Fitting model parameters to data amounts to finding a minimum of fitting error over the training set. It is crucial that the minimum is found with a sufficient accuracy. Imprecise solutions are not only suboptimal in the fit quality but can also lose important properties such as unbiasedness. Methods whose constituent part is the comparison of evaluations on a verification set such as those of Sects. 4.7.1 and 4.8 can only work if individual evaluations are done with the same precision.

For linear models, explicit solutions of this minimization problem are known. With nonlinear models such as neural networks, iterative numeric methods have to be used. For this case, extensive theory of numeric optimization is available, depending on the class of optimization problem. Most methods use the gradient of the error function.

There are three basic classes, with growing generality:

(A) *Convex problems, where the objective function to be minimized is convex in parameters.* For this problem class, fast converging algorithms have been developed. These are called second-order methods, an example of which is the conjugate gradient method. They are considerably superior to the first-order methods using a simple gradient descent.

(B) *Problems with a single local minimum.* These problems may exhibit non-convex, but descending regions of the objective function. However, there is always a convex region around the minimum. For these problems, there are no guarantees for convergence speed. Nevertheless, the second-order methods can also be used, giving the advantage of fast convergence in the convex regions.

(C) *Problems with multiple local minima.* For this problems class, solution methods with convergence guarantees exist only for problems with very small numbers of parameters (usually below ten). Such problem sizes are hardly relevant for Data Science tasks. The methods used for these problems are essentially heuristic, attempting to find a good, although suboptimal solution. Their success heavily depends on the particular problem.

Parameter fitting of nonlinear models, in particular neural networks, is generally of type (C), but can, in individual cases, be simpler, i.e., of type (B) or even (A). It is known that every layered network possesses combinatorial invariances leading to a large number of equivalent minima. It is then sufficient to reach any of these minima. More serious is the extent of nonlinearities: Mappings with strong nonlinearities are more probable to produce non-convexities, or even multiple, non-equivalent minima of the error function.

In practice, there is acceptable success record of methods adequate for class (B), sometimes supplemented with heuristics such as momentum terms accounting for possibly encountering class (C).

5.7 Comprehension Check

1. What is the reason for the poor convergence of the first-order gradient descent with convex minimization problems?
2. What are the memory requirements of the conjugate gradient method if the parameter vector over which the minimization takes place is of size P?
3. Are there practical numerical optimization methods with convergence guarantees for large minimization problems with multiple minima?
4. What is the idea behind the momentum term (with regard to the surface of function to be minimized)?
5. What is the basic problem in using the momentum term?

References

1. Åström KJ, Wittenmark B (1989) Adaptive control. Addison-Wesley series in electrical and computer engineering, Addison-Wesley, Reading
2. Bermeitinger B, Hrycej T, Handschuh S (2019) Representational capacity of deep neural networks: a computing study. In: Proceedings of the 11th international joint conference on knowledge discovery, knowledge engineering and knowledge management, vol 1. KDIR, SCITEPRESS. Science and Technology Publications, Vienna, Austria, pp 532–538. https://doi.org/10.5220/0008364305320538
3. Devlin J, Chang MW, Lee K, Toutanova K (2019) BERT: pre-training of deep bidirectional transformers for language understanding. In: Proceedings of the 2019 conference of the North American Chapter of the Association for Computational Linguistics: human language technologies, vol 1 (Long and Short Papers), Association for Computational Linguistics, Minneapolis, Minnesota, pp 4171–4186. https://doi.org/10.18653/v1/N19-1423
4. Dikin II (1967) Iterative solution of problems of linear and quadratic programming. Dokl Akad Nauk SSSR 174(4):747–748. http://mi.mathnet.ru/dan33112
5. Duchi J, Hazan E, Singer Y (2011) Adaptive subgradient methods for online learning and stochastic optimization. J Mach Learn Res 12(null):2121–2159

6. Fletcher R, Reeves CM (1964) Function minimization by conjugate gradients. Comput J 7(2):149–154. https://doi.org/10.1093/comjnl/7.2.149
7. Goodfellow I, Bengio Y, Courville A (2016) Deep learning: adaptive computation and machine learning. The MIT Press, Cambridge
8. Hinton G (2012) Neural networks for machine learning. http://www.cs.toronto.edu/~tijmen/csc321/slides/lecture_slides_lec6.pdf
9. Holland JH (1992) Adaptation in natural and artificial systems: an introductory analysis with applications to biology, control, and artificial intelligence, 1st edn. Complex Adaptive Systems, MIT Press, Cambridge, Mass
10. Kingma DP, Ba J (2015) Adam: a method for stochastic optimization. ICLR. http://arxiv.org/abs/1412.6980
11. Kirkpatrick S, Gelatt CD, Vecchi MP (1983) Optimization by simulated annealing. Science 220(4598):671–680. https://doi.org/10.1126/science.220.4598.671
12. Levenberg K (1944) A method for the solution of certain non-linear problems in least squares. Quart Appl Math 2(2):164–168. https://doi.org/10.1090/qam/10666
13. Marquardt DW (1963) An algorithm for least-squares estimation of nonlinear parameters. J Soc Indus Appl Math 11(2):431–441. https://doi.org/10.1137/0111030
14. Metropolis N, Rosenbluth AW, Rosenbluth MN, Teller AH, Teller E (1953) Equation of state calculations by fast computing machines. J Chem Phys 21(6):1087–1092. https://doi.org/10.1063/1.1699114
15. Mockus J, Eddy W, Reklaitis G (1997) Bayesian heuristic approach to discrete and global optimization: algorithms, visualization, software, and applications. Nonconvex optimization and its applications. Springer, Berlin. https://doi.org/10.1007/978-1-4757-2627-5
16. Niewiadomski S (2013) Filter handbook: a practical design guide. Newnes, Amsterdam
17. Polak E, Ribière G (1969) Note sur la convergence de méthodes de directions conjuguées. RIRO 3(16):35–43. https://doi.org/10.1051/m2an/196903R100351
18. Potra FA, Wright SJ (2000) Interior-point methods. J Comput Appl Math 124(1):281–302. https://doi.org/10.1016/S0377-0427(00)00433-7
19. Press WH, Teukolsky SA, Vetterling WT, Flannery BP (1992) Numerical recipes in C: the art of scientific computing, 2nd edn. Cambridge University Press, Cambridge
20. Rinnooy Kan AHG, Timmer GT (1987) Stochastic global optimization methods part I: clustering methods. Math Program 39(1):27–56. https://doi.org/10.1007/BF02592070
21. Robbins H, Monro S (1951) A stochastic approximation method. Ann Math Statist 22(3):400–407. https://doi.org/10.1214/aoms/1177729586
22. Rosenblatt F (1957) The perceptron—a perceiving and recognizing automaton. Cornell Aeronautical Laboratory
23. Simonyan K, Zisserman A (2015) Very deep convolutional networks for large-scale image recognition. ICLR
24. Widrow B, Hoff ME (1960) Adaptive switching circuits. Technical report 1553-1, Solid-State Electronics Laboratory, Stanford
25. Yarlagadda RKR (2010) Analog and digital signals and systems. Springer, New York

Part II
Applications

The diversity of DS applications makes even a rough method overview infeasible. However, some applications have strongly gained momentum in recent years. Two of them can be viewed as representatives of the current state of AI with a particular impact in applications:

- *Natural Language Processing* (NLP) and
- *Computer Vision* (CV)

Another particularity of these two applications is that they deal with objects (sentences or images) whose characterization by a vector of features is not straightforward. Moreover, individual parts of these objects can only be reliably interpreted in their respective context (e.g., words in a sentence or patches in an image). Considering and exploiting this context is a particular challenge of these applications.

This is why these two fields are mentioned separately in the following two chapters. The presentation focuses on mathematical principles that are broadly used and, as far as possible, justified by more than merely computing experience. There is clearly no claim of completeness in coverage of the methods, whose scope is very broad and depending on the characteristics of the particular application.

Specific Problems of Natural Language Processing

6

Natural Language Processing (NLP) is a field going far beyond the scope of DS. It combines computer science and linguistics by using computational techniques in order to process, understand, and produce human language content [12]. The application areas are very broad and range from tasks such as extracting information from text, translating or generating text, to full conversational systems (e.g., chatbots or personal assistants) that can interact with their users [6, 18, 19, 21].

NLP also covers the area of spoken language processing, where the language is not given in its written form, but as an audio signal or voice recording of one or more speakers. A natural approach to process spoken language is to utilize so-called *speech-to-text systems* which can convert spoken language into text transcriptions and then apply the common methods for text processing [4]. However, direct processing of the audio signal offers the possibility of also capturing implicit features such as pitch or energy, which can be used to detect different voices or even emotions. In this case, a number of special aspects of acoustic signal processing must be taken into account [32]. Spoken language processing is outside the scope of this book. Nevertheless it is worth noting that some of the mathematical principles and architectures presented in the following chapters (e.g., the Transformer architecture) have already found their way into this area as well.

Processing text incorporates linguistic concepts such as syntax, semantics, or grammar, some of which can be formalized. In their formalized form, they belong rather to the domain of discrete mathematics or mathematical logic. However, some approaches are clearly data driven. A part of them can be summarized under the concept of *corpus-based semantics* [3]. This discipline extracts the semantic meaning of the text by statistical means from a corpus (i.e., a large collection) of texts. It is this subfield of NLP that is addressed in the following sections. The topics of the sections are selected according to the extent to which mathematical principles are used. Such selection cannot, of course, cover the complete domain of NLP—less formalizable aspects are omitted.

© The Author(s), under exclusive license to Springer Nature Switzerland AG 2023   167
T. Hrycej et al., *Mathematical Foundations of Data Science*, Texts in Computer Science, https://doi.org/10.1007/978-3-031-19074-2_6

6.1 Word Embeddings

For a language processing system based on a real-valued representation (such as neural networks), all input data must first be transformed into a numerical representation. Unlike numerical data or data that already have an inherent numerical structure, such as images where each pixel can be specified by a vector of values for each color channel (e.g., three values for RGB), sentences have no such obvious scheme.

A major task in NLP is therefore the transformation of text into a sequence of real-valued vectors. The vectors are usually determined at word level. This is done by assigning a vector to each word position in the text sequence. However, there are also some recent approaches which reduce the granularity of the vectors to single word parts (so-called *subwords* or *wordpieces*) [33]. In order not to go beyond the scope of this book, we will focus on vector representations for discrete words.

A simple vector representation of a given word can be created by constructing a vocabulary V which contains a list of all potential words. Since it is not feasible to determine all possible words in advance, the vocabulary is often constructed with the most frequent words up to a predefined vocabulary size v. In addition, a special OOV token is often used to represent all other words that are not in the vocabulary: $V = [w_1, \ldots, w_{v-1}, \text{OOV}]$.

By assigning each word w_i to its index i in the vocabulary, we can transform these numbers into so-called *one hot encoding* vectors [31]. Such vectors solely consist of 0 values and a single value of 1 at the particular position i of w_i in the vocabulary. In this way, any word $w_i \in V$ can be transformed into a vector $t_i = [o_{i1}, \ldots, o_{iv}]$ of dimension v which has the following properties:

- $o_{ij} = 1$ for $i = j$; and
- $o_{ij} = 0$ for $i \neq j$

Note that the representational power of these vectors is very limited. They solely capture information about a word's position in the vocabulary which can be arbitrary. Furthermore, the vector representations are sparse vectors of high dimensionality determined by the vocabulary size v. They allow no meaningful mathematical definition of similarity. Since all such vectors are mutually orthogonal, the vector product of any two vectors representing different words is zero. For a representation to be useful for semantic processing, we aim for a dense representation of real-valued vectors with low dimensionality that can express the similarity between words in different contexts [31].

Word vectors mostly serve a higher goal through providing an initial representation within a larger mathematical model to solve a certain task. Thus, they are usually not evaluated themselves, but only in terms of how well the task could be solved by them. This makes it very difficult to give a general guideline about what properties word vectors should have and how to assess their quality.

However, commonly observed properties of word vectors that led to good downstream performance have shown that they

- capture the syntactic and semantic information of a word;
- preserve syntactic and semantic similarity, meaning that similar words are to be represented by similar vectors (e.g., synonyms) [25]; and
- model their syntactic and semantic usage in different contexts (e.g., polysemy) [23].

The latter addresses the very important concept of *context*, where two words are assumed to have similar meanings when they appear in similar contexts. Consider the words "beagle" and "terrier". They may often co-occur with words such as "bark", "leash," or "hunt". Both words are nouns and therefore may often appear as subjects or objects in a sentence. Because of their similarity, these words are expected to have a small distance between their corresponding word vectors in the vector space.

There exist many possibilities to create word vectors. They can be generated using matrix factorization methods of a global co-occurrence matrix which consists of pair-wise count values on how many times two words appear together in a specific context window [22]. In many cases, however, word vectors are learned hidden representations within a larger neural network. In this context, they are often referred to as *word embeddings* [31]. They typically form the very first layer of a series of stacked layers in a neural network architecture. During training, word embeddings are learned along with all other model weights, optimized for the task the network is intended to solve (e.g., text classification) or a set of general pretraining tasks that foster good task-independent representations.

When used as the first layer in a neural network, often the desired properties of word embeddings are not formed until the subsequent layers. This is especially true when the input sequence does not process words but subwords. In this case, the semantics of a word would result from the combined processing of its subwords further down in the neural network [5]. For this reason, and despite their simplicity and inherent shortcomings, vocabulary-based one hot encoding vectors are still predominantly used as input vector representations in neural architectures to create word embeddings with lower dimensionality.

In the simplest case, this can be done by a single linear transformation described by a $(v \times d)$ weight matrix M which is learned during the training process. Thereby, the embedding dimension d is usually set to a value that is significantly smaller than the vocabulary size v, e.g., 300. For each word in the vocabulary, its embedding vector x_i corresponds to the i-th row in M.

This transformation can be described as follows:

$$x_i = Mt_i \tag{6.1}$$

where t_i is the one hot encoded vector for word w_i. If the NLP system has the form of a neural network, this process of mapping vocabulary-based indices to vector representations is often referred to as an *embedding layer* [31].

In addition to word embeddings that capture semantic and syntactic context, information about the order of words in the sentence can be passed as extended input into the network. This information is especially useful when the mathematical operators in the network are not position aware, meaning that they process each element in the input sequence in the same way.

The positional information can be supplied in the form of an additional learned vector p_j (or *position embedding*) which has the same dimensionality as the semantic embedding vector x_i.

$$x_i^* = x_i + p_j \tag{6.2}$$

In the simplest case, p_j can be constructed analogously to x_i by having its own embedding layer. Instead of considering word positions in the vocabulary, this layer uses the absolute word positions in the input sequence to learn a vector representation associated with each input position j. As a result, same words that occur in different positions in the input sentence will have different embedding vectors x_i^* [8].

It is intuitively clear that in the most cases, it is the relative position between two words that is decisive for the semantics, rather than the absolute one. One possible approach is to rely on finding out the relative position by learning. An alternative is to express the relative position directly. It is clear that this cannot be done for a single word but only for a pair of them. For this reason, processing relative positions usually resembles the mathematical definition of a similarity concept between word pairs, which is discussed in the following chapter.

6.2 Semantic Similarity

Processing natural language with a focus on semantics requires the possibility to recognize which semantic representations are close to each other. For example, synonyms are expected to have such close, or almost identical, semantic representations.

To resolve ambiguities (which are very frequent in natural language if observed on the word level), it is helpful to pick up words or phrases that can contribute to such disambiguation. In other words, parts of the sentence that are relevant for the semantics of a given word have to be identified. For this goal, similarity of representation is a useful tool.

When words are represented as numeric vectors, mathematical definitions of similarity can be used. A straightforward way to assess vector similarity is a vector product. If word A is encoded by vector v_A and word B by vector v_B, their similarity can be measured by their vector product:

$$w_{AB} = v_A{}' v_B. \tag{6.3}$$

A better insight to the interpretation of such weights is provided by the concept of *projection*. Every vector can be decomposed to (1) a part showing in a predefined direction and (2) a part showing to a direction orthogonal to the predefined one. This is depicted in Fig. 6.1.

Here, vector v_B has a part showing the direction of vector v_A, which is denoted as $v_{B|A}$. It is called the projection of v_B onto v_A. The more similar the directions v_B and v_A are, the longer is the projected part $v_{B|A}$. If the direction of both vectors was orthogonal, the projection would be zero.

Fig. 6.1 Orthonormal projection of vector v_B onto vector v_A

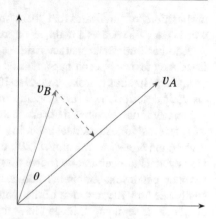

The algebraic description of this relation is

$$v_{B|A} = \frac{v_A v_A'}{\|v_A\|} v_B = \frac{v_A v_A'}{v_A' v_A} v_B = v_A \frac{v_A' v_B}{\|v_A\|} = \frac{v_A}{\|v_A\|} w_{AB} \qquad (6.4)$$

The first two terms correspond to the usual form of a projection operator. The last term is in the form of a normalized vector v_A weighted by the weight w_{AB}. For a lexical pattern (e.g., a word) B, $v_{B|A}$ shows its "part" that is related to word A.

A derived measure is the normalized cosine similarity which quantifies $v_{B|A}$ between -1 and 1. It is very popular because it is independent of the length of the two vectors v_A and v_B. Cosine similarity corresponds to the cosine of the included angle θ. A value of -1 indicates the exact opposite, with both vectors pointing in opposite directions, while 1 indicates maximum similarity, with both vectors having identical directions. The value 0 represents orthogonality of both vectors.

$$w_{AB}^{\star} = \frac{v_A' v_B}{\|v_A\| \|v_B\|} = \frac{v_A' v_B}{v_A' v_A v_B v_B'} \qquad (6.5)$$

To summarize, w_{AB} and $w_{AB}^{\star}$ express how similar the words A and B are. Seeking the semantics of word A, all other words B can be scanned for similarity to help to disambiguate the semantics of A. These similarity measures can be used as a weight of relevance of B.

6.3 Recurrent Versus Sequence Processing Approaches

An important concept in NLP is the notion of *context*, which means that words can have different meanings in different linguistic relationships. As an example, consider the word "bark", which can refer to a tree's outer layer or to the sound that a dog makes. However, its meaning can be inferred by the context in which it is used. For example, in the sentence "The dog likes to scratch the bark of the tree.", the preposition "of" indicates that "bark" must be a part of the tree. In this context, the work "bark" is used as a noun, as opposed to the verb "to bark".

Another important characteristic of language processing is the sequential text flow. Any text or speech has a natural order in which the words are written, read, or spoken. In simple tasks, e.g., classifying products into specific categories based on their descriptions, it may be appropriate to ignore this sequential information, especially if one assumes that certain keywords are good predictors on their own. In this case, a simple numerical encoding of the set of words in the product description, a so-called *bag-of-words model* [20], could be sufficient as an input representation. However, it has been clearly shown that considering the natural order of the text leads to much better text comprehension, especially in more complex tasks [34]. We can easily see how the meaning of the word "bark" becomes ambiguous as soon as we consider the sentence "The dog likes to scratch the bark of the tree" as an unordered set of its words { "the", "dog", "likes", "to", "bark", "scratch", "of", "tree" }.

All in all, the semantics of particular text positions (subwords, word, or phrases) should be determined not only by itself but also through other positions. Furthermore, the order of the positions should be taken into account.

This suggests the possibility to view semantics of a certain position as a function of a text window around this position. When processing long sequences or sequences that are still being generated (e.g., in *text generation* tasks), we usually only have access to the previous positions that have already been processed. This is usually denoted as *left-to-right* processing. So in the above sentence "The dog likes to scratch the bark of the tree", the semantics of "bark" would be a function of, e.g., "The dog likes to scratch the bark". However, if the complete text is available at the time of the analysis, also the subsequent positions can be included and contribute to the disambiguation of semantics.

This trivial example suggests the problems that can be expected to occur:

- It is difficult to determine the size of the sequence capturing the genuine meaning.
- The relevance of individual words in the sequence strongly varies.
- Assigning a particular role to a position in the sequence is not possible since the positions are frequently interchangeable within the framework of the language grammar.

Because of these particular properties of NLP, it is important to find an appropriate representation for the natural language flow. This has been a research topic for decades, with many different approaches [16].

Generally speaking, text processing can be viewed as an instance of dynamic system modeling as discussed in Sect. 2.3. This suggests which general representation options are available for the modeling.

Viewing the semantics of a text position as a static function of its predecessors corresponds to a mapping whose input is a text sequence up to the current one, the output being the semantics of the current position. Treating a text sequence in this way is analogous to the FIR of Sect. 2.3. A typical neural network implementation is feedforward (see Sect. 3.2).

From the viewpoint of the dynamic system theory, it is not unexpected that an alternative approach to text processing is also available. It is an analogy to the IIR. In this approach, a recurrent model is used whose internal state depends on its own past values. A typical implementation is by a neural network of the type described in Sect. 3.3.

Although both processing alternatives go beyond the basic variants of feedforward and feedback networks, they share some of their favorable and unfavorable properties:

- Direct processing of the text sequence requires input vectors of considerable size (up to hundreds of words). On the other hand, their feedforward character avoids problems of instability or oscillatory behavior.
- Recurrent models receive input vectors of moderate size corresponding to a word or its immediate neighborhood. The representation of past inputs is expected to be formed in the internal state (which is generally true for dynamic models of this type). However, there is a substantial danger of instabilities or oscillations that are not always easy to detect and eliminate.

In the following sections, prominent examples of both approaches will be briefly presented: the recurrent approach in Sect. 6.4 and the sequential approach in Sect. 6.5 through Sect. 6.7.

6.4   Recurrent Neural Networks

While sequence processing was a common practice for the first attempts, the recurrent approach became dominant for a long time. This was mainly due to so-called *Recurrent Neural Networks* (RNNs), which are theoretically able to process sequences of arbitrary length. However, they became successful only after some modifications of RNNs had been developed, which, through various adjustments, were actually able to handle long-ranging dependencies between words in practice. Probably the most prominent representative of this genus are Long Short-Term Memory Neural Networks (LSTMs) which were introduced by Hochreiter and Schmidhuber [13].

Their development made RNNs to important building blocks everywhere where input sequences of arbitrary length need to be processed or transformed into new output sequences. This is especially true for natural language and audio processing applications, where RNNs can be used to generate text from a given text sample, translate texts into another language, or generate transcriptions from an audio recording [7,24,29]. RNNs process each word of the sequence in a particular order, usually from left-to-right in a *unidirectional* fashion. Thereby, the internal or *hidden* state h_t, which is supposed to represent the semantics of the words already processed, is continuously updated for each position t while producing an output.

Let x_t denote the vector representation of a particular word in the input sequence, y_t the vector output at that particular position, and h_t the vector of mentioned hidden states. Then, RNNs are very similar to the definition of discrete linear systems in state-space form, as described in (2.73) and (2.74) of Sect. 2.3 [7]:

$$h_t = H\,(h_{t-1}, x_{t-1}) \tag{6.6}$$

$$y_t = Y\,(h_t, x_t) \tag{6.7}$$

(In the recurrent network community, it is usual to denote the input as x, in contrast to the dynamic system literature referred in Sect. 2.3, where x is the state and u the input. In addition, indices in 6.6 have been shifted in order to view the hidden states as a result of past hidden states and inputs.)

However, there are differences in the mathematical formulation. First of all, in RNNs (see 6.8 and 6.9), H is typically represented by a parameterized nonlinear function instead of a linear function. This function can be arbitrary complex. Secondly, the internal state h_t is defined as directly dependent on the current input x_t as opposed to its previous value x_{t-1}. Finally, x_t was removed from the equation used to calculate the output y_t. In linear dynamical systems, this dependence of y_t on x_t indicates the "feedthrough" (the non-dynamic influence) of the input on the output. In RNNs, however, this influence is retained by y_t depending on the hidden state h_t, which in turn depends on x_t. In summary, both representations can be considered equivalent except for the nonlinearity, with only differences in notation and indexing:

$$h_t = H\,(h_{t-1}, x_t) \tag{6.8}$$

$$y_t = Y\,(h_t) \tag{6.9}$$

It is worth mentioning that in RNNs, there is often no notion of an external output besides the hidden state that is passed into the computation of the next state. In this case, Y takes on the identity function $id\,(x) = x$ and the hidden state h_t is thus denoted as the output y_t which is then used to calculate the next hidden state h_{t+1} [13].

The repetitive or *recurrent* calculations of h in RNNs, which in turn are based on previous calculations of h, can be visualized as a successive chain of calculation blocks as depicted in Fig. 6.2. Here, a block is determined by the two calculations of h and y and is usually referred to as as a *cell* (or *memory cell* in LSTMs). The different cells in the chain do not represent different cell instances with their own learnable weights, but visualize the same calculation continuously over time for different inputs (e.g., words in a sentence).

In theory, the recurrent structure allows to process sequences of arbitrary length, enabling the model to access previous information in a sentence up to the current word. This allows the semantics of a word to be disambiguated by words that do not necessarily reside in the immediate vicinity. In practice, however, it has been shown that simple RNNs lead to vanishing gradients over long-ranging dependencies. This is a serious obstacle for learning by numerical optimization methods.

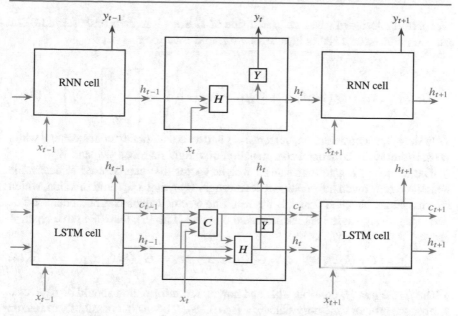

Fig. 6.2 Comparison between RNN and LSTM cells—the cells in the middle reveal the respective calculation

Vanishing gradients denote a problem that occurs if the derivatives of the error function are chained through a nested structure such as a layered neural network. Nested nonlinear activation functions lead to diminishing derivative size, in particular if saturation segments of the functions' range are involved. In RNNs, the derivatives are recursively passed through the same neural network (by feedback), resulting in a particularly long chain that can cause the gradients to vanish as they become smaller and smaller. As a consequence, RNNs are known to have difficulty learning long-range dependencies [14].

As a result, several variants of RNNs emerged that address this problem of RNNs' short-term memory. Often, they introduce additional states that can be controlled by the network by so-called *gates* that determine which and how much information from the states should be updated. One example is LSTMs [13], which use an additional state c that fosters long-term memory by being updated only selectively. As a result, the single RNN state vector h from 6.8 is replaced by two state vectors, h and c:

$$h_t = H\left(h_{t-1}, x_t, c_t\right) \tag{6.10}$$

$$c_t = C\left(h_{t-1}, x_t, c_{t-1}\right) \tag{6.11}$$

$$y_t = h_t \tag{6.12}$$

In LSTMs, Y is represented by the identity function where the hidden state h_t is returned as the output. For updating the memory state c_t and controlling the output

y_t, specially designed gates are used inside of H and C. In their basic form, LSTM gates G_i consist of a neural layer with a sigmoid function σ:

$$G_i\left(h_{t-1}, x_t\right) = \sigma\left(W_{h_i} h_{t-1} + W_{x_i} x_t + b_i\right) = \sigma\left(W_i \begin{bmatrix} h_{t-1} \\ \vdots \\ x_t \end{bmatrix} + b_i\right) \quad (6.13)$$

In the latter representation, vectors h_{t-1} and x_t are vertically concatenated, while W_i substitutes the horizontal concatenation of weight matrices W_{h_i} and W_{x_i}.

The output of a gate represents a weight vector that can be used to determine which and how much information to let through. By using a sigmoid function, which returns values between 0 and 1, the resulting vector values can be understood as percentage weights. In LSTMs presented by Gers et al. [9], there are three types of gates:

$$f_t = G_f\left(h_{t-1}, x_t\right) \quad i_t = G_i\left(h_{t-1}, x_t\right) \quad o_t = G_o\left(h_{t-1}, x_t\right) \quad (6.14)$$

• The *forget gate* f_t controls what and how much information should be neglected from the previous memory state c_{t-1} (see 6.16). It opens the possibility to free up old context information that is no longer relevant.
• The *input gate* gate i_t controls what and how much information from the input is written to the memory state c_t (see 6.16). It gives a sense of how important certain aspects of the current input word are for the current context.
• The *output gate* o_t controls what and how much output information is read from the memory state c_t (see 6.15). This selected information, passed to the next state computation, can help process the next word in the sequence.

$$h_t = y_t = H\left(h_{t-1}, x_t, c_t\right) = o_t \circ \tanh\left(c_t\right) \quad (6.15)$$

$$c_t = C\left(h_{t-1}, x_t, c_{t-1}\right) = f_t \circ c_{t-1} + i_t \circ C^\star\left(h_{t-1}, x_t\right) \quad (6.16)$$

where $\circ$ is the element-wise or *Hadamard product* operator and $C^\star$ is defined as follows:

$$C^\star\left(h_{t-1}, x_t\right) = \tanh\left(W_{h_c} h_{t-1} + W_{x_c} x_t + b_c\right) \quad (6.17)$$

The gate values are produced by a sigmoid function, which determines how much information should be written or read from the memory state c on a scale from 0 (no pass-through) to 1 (full pass-through). The actual value to be written or read is passed through the tanh (tangens hyperbolicus, see Sect. 3.1) function which is close to a linear function in a close environment of zero argument and saturates at -1 and 1. It produces negative and positive values between -1 and 1. In other words, the state vectors h and c are updated by saturated, weighted functions of the input and the state [10].

Using saturated activation functions such as sigmoid and tanh is a usual nonlinearity representation in neural networks. It prevents the activation values from growing arbitrarily if passed through multiple layers. For recurrent networks, it additionally helps to avoid infinite growth (up to numeric overflow) if the network parameters imply instability. However, as briefly discussed in Sect. 3.3, the instability is not genuinely removed but only "shadowed" by saturation (see Fig. 3.10).

This mathematical formulation of an LSTM is intended to give a rough overview of how LSTMs work. It represents one of the first LSTM versions in a series of improvements and changes, which cannot all be covered in this book. For example, further corrections have been made by the authors, such as the integration of *peephole connections*, whereby the gates can also be controlled by the memory state c in addition to h and x. Another well-known variant is GRUs [7], which have a simpler structure. Here, only one internal state h_t is used, while a new state $\tilde{h}_t$ is proposed every time a word is processed. An *update state* decides whether the current state h_t should be updated with the new hidden state $\tilde{h}_t$ and a *reset gate* determines whether the previous state h_{t-1} is ignored when computing $\tilde{h}_t$.

So far, we have encountered RNNs as an important representative of unidirectional language models where the input sequence is traversed in a certain order (here: left-to-right). In this case, only information from previous positions can be accessed while processing a position in the input sequence. If the sequence consists of a complete sentence, this restriction can lead to the fact that important words later on in the sentence, which contribute to the semantics of the current word, are not taken into account. Therefore, it is not surprising that so-called *bidirectional* language models have also been developed. These are no longer subject to this limitation and are able to access past positions as well as subsequent positions in a given sequence. In the field of recurrent processing, this can be solved with an additional RNN, which processes the sequence in negative time direction (*right-to-left*). By combining a left-to-right RNN with such a right-to-left RNN, the resulting model can be trained in both directions simultaneously. These bidirectional models have outperformed unidirectional RNNs, especially in classification and regression tasks [27].

Despite their long success, recurrent models still suffer from the unfavorable mathematical properties already elaborated in Sect. 3.3. These factors have therefore recently led to the revival of sequence processing. A prominent example of a bidirectional sequence processing approach is the system BERT [8] with an excellent success record throughout a broad scope of NLP applications. Underlying mathematical principles of some of its innovative features will be presented in the following sections.

6.5   Attention Mechanism

The revival of simultaneous text sequence processing brought about the problem known from FIR filters: the length of the filter input is substantially larger than that of a IIR filter and so is the text sequence input. Since many words in such a long sequence may not directly contribute to the meaning of a single selected word, it makes sense to focus on those where the chance for relevance is high. This can be viewed as an analogy to the cognitive concept of attention—we pay attention to what we consider to be probably important.

An *attention mechanism* can be used in order to focus on relevant elements during sequence processing. In contrast to pure FIRs, the attention mechanism does not always assume that the output depends only on a fixed-size region of past input values, but can process the entire input sequence including subsequent words at once.

The goal of the attention mechanism is to learn important regions of the full input sequence. This is done by defining an attention distribution over all positions in the input sequence where high values indicate positions with high importance. This distribution is then used to perform a weighted selection over the input sequence [2, 17].

Let $v = [v_1, \ldots, v_n]$ represent a sequence of input vectors. Then, the output of an attention mechanism (often called *context vector*) c is defined by the weighted sum of the inputs $c = \sum_{i=1}^{n} \alpha_i v_i$. Instead of a single output, an attention mechanism usually calculates a sequence of context vectors that are used in different contexts. Hence, each context vector c_j has its individual distribution over the input sequence thus focusing on different areas of the input:

$$c_j = \sum_{i=1}^{n} \alpha_{ji} v_i \tag{6.18}$$

An attention weight α_{ji} thereby defines the importance of an input vector at position i with respect to the output position j. Together, all attention weights form a distribution of relevance over the input sequence. One way to achieve this is with help of the softmax function:

$$\alpha_{ji} = \frac{e^{e_{ji}}}{\sum_{k=1}^{n} e^{e_{jk}}} \tag{6.19}$$

where

$$e_{ji} = f(q_j, k_i) \tag{6.20}$$

represents an *alignment model* that defines how well the inputs around position i and the output at position j match [2]. The alignment model could be any similarity measure of two vectors q_j and k_i as described in Sect. 6.2, or a more complex model like a feedforward neural network. To account for measures with different value ranges (e.g., the similarity measures of Sect. 6.2 can return both positive and negative values), the softmax function ensures that the resulting weights are positive numbers between 0 and 1 that sum up to unity. This corresponds to a probability distribution with n possible outcomes (for each position in the input). Usually, a simple matrix product is used to calculate the similarity between q_j and k_i, often scaled by the embedding dimension k:

$$e_{ji} = \frac{q_j' k_i}{\sqrt{k}} \tag{6.21}$$

As a result of the properties of the vector product, this similarity metric leads to either positive or negative sign. Negative sign with a large absolute similarity value occurs if q_j and k_i show in opposite directions. It is not simple to decide what is "less similar":

- the opposite sign case (possibly corresponding to opposite meaning which may point to the same semantic category); or
- the similarity close to zero, that is, orthogonal vectors (possibly corresponding to meanings that have nothing to do with each other).

Taking attention weights proportional to the exponential function of the similarity measure classifies the opposite sign case as less similar than the orthogonal ones (similarity close to zero). This should be kept in mind if softmax weighting concept is used.

Intuitively, the attention mechanism for a single output c_j can be seen as a kind of fuzzy database search where database entries are stored as key-value pairs. It is helpful to think of the keys as something that has an encoded meaning, such as named keys used in a *JSON*-object structure: {"key": value}. For example, consider the following database which consists of food prices per kilogram, e.g., {"oranges": 2}, {"walnuts": 20}, and {"nutmeg": 30}. When receiving a database query (here: "peanuts"), the goal is to find entries whose keys are similar to the query. The higher the similarity, the more relevant we would consider the item to be to the query. Similarities to the queries can be calculated using an alignment model f. Let's assume the following similarities, on a scale from -1 to 1:

$$f\ (\text{"peanuts"}, \text{"oranges"}) = -0.7$$
$$f\ (\text{"peanuts"}, \text{"walnuts"}) = 0.8$$
$$f\ (\text{"peanuts"}, \text{"nutmeg"}) = 0.6$$

Once the relevant items are identified, we would use their values for the final result. In the attention mechanism, we would not return a list of results, but a soft-selection of the input values using a weighted sum. In our example, this corresponds to an average kilo price for items that are similar to "peanuts":

$$\alpha\ (\text{"peanuts"}, \text{"oranges"}) = \frac{e^{-0.7}}{e^{-0.7} + e^{0.8} + e^{0.6}} \approx 0.11$$

$$\alpha\ (\text{"peanuts"}, \text{"walnuts"}) = \frac{e^{0.8}}{e^{-0.7} + e^{0.8} + e^{0.6}} \approx 0.49$$

$$\alpha\ (\text{"peanuts"}, \text{"nutmeg"}) = \frac{e^{0.6}}{e^{-0.7} + e^{0.8} + e^{0.6}} \approx 0.40$$

$$c = \alpha\ (\text{"peanuts"}, \text{"oranges"})\ v\ (\text{"oranges"})$$
$$+\ \alpha\ (\text{"peanuts"}, \text{"walnuts"})\ v\ (\text{"walnuts"})$$
$$+\ \alpha\ (\text{"peanuts"}, \text{"nutmeg"})\ v\ (\text{"nutmeg"})$$
$$= 0.11 \cdot 2 + 0.49 \cdot 20 + 0.40 \cdot 30$$
$$= 22.02$$

Because of this analogy, the vectors for the alignment model $f(q_j, k_i)$ are often referred to as *query vectors* q_j and *key vectors* k_i, while the vectors of the input sequence v_i are often called *value vectors* [30].

6.6   Autocoding and Its Modification

If the purpose of an algorithm is not a particular application but rather delivering an appropriate representation for broad scope of NLP tasks, a possible method is to look for a good encoding in the sense of Sect. 2.5.2. Optimal encoding consists in finding a transformation of input vectors to a low-dimensional space such that an inverse transformation back to the original space fits the input in an optimal way.

BERT [8] modifies this principle by trying to fit only a part of the reconstructed input. This principle is called *masking*. The BERT mapping through multiple Transformers is trained to predict parts of the input sequence from which some randomly selected words are left unknown. It can be expected that this task is easier than predicting the complete text sequence, alleviating the huge computational burden for training. A precondition for this principle, as for every encoding, is that intermediary representations in the system have lower dimension than input and output. Otherwise, an identity mapping would be the optimum solution.

6.7   Transformer Encoder

There is a broad variety of architectures for language processing, and it is beyond the scope of this book to discuss even a small representative selection of them. However, it is certainly interesting to show what such an architecture may look like in more detail. To present a prominent example, the successful concept of the so-called *Transformer* architecture [30] has been chosen which is build upon the already presented concept of attention in Sect. 6.5.

The first NLP attempts that adopted attention for sequence processing utilized the attention mechanism in combination with recurrent approaches as a way to process sequences of hidden states that have been produced with RNNs such as Long Short-Term Memory Neural Networks (LSTMs). Thereby, instead of using only the last hidden state as a meaningful representation of the full sentence, attention mechanism allows to selectively access information from all hidden states by focusing on the most relevant positions only. Another advantage of using attention mechanisms over recurrent methods is that they shorten the paths of forward and backward passes in the network, leading to better learning of long-range dependencies [2, 30].

In 2017, Vaswani et al. [30] introduced the Transformer architecture which uses a novel encoding mechanism that relies solely on attention mechanisms without the need for recurrence. In this chapter, we will focus on one of the key elements of the Transformer, the *Transformer encoder*.

Intuitively, we can formulate the goal of such a Transformer encoder as transforming initial input vector representations x_i into new context-aware vector representations y_i. While the inputs x_i may (or may not) capture general syntactic or semantic information about a word, the output vectors y_i are expected to further model their syntactic and semantic use within the given sentence, thus making them aware of their context [35]. This property is mainly achieved by using multiple layers of at-

Fig. 6.3 Transformer encoder

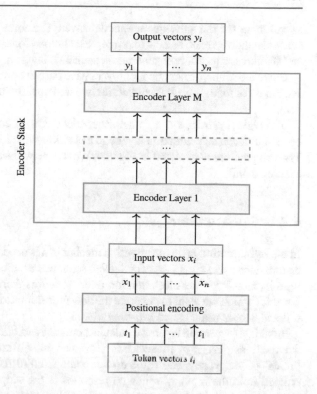

tention mechanism that process the whole input sequence in a bidirectional fashion. Thereby, the output representations y_i are learned by jointly observing context from left $[0, \ldots, i-1]$ and right positions $[i+1, \ldots, n]$. By considering both left and right context, Transformer encoders have been successfully used by the system BERT [8] to produce *contextualized* representations when trained on large text corpora, using a number of different pretraining tasks. These contextualized representations differ from other context-independent representations in a way that they allow the model to be interpreted differently in different contexts [23,35]. As a result, they are very well suited to be used for a broad scale of NLP applications.

The main structure of a Transformer encoder is visualized in Fig. 6.3.

When processing text, the inputs for the first encoder layer x_i usually correspond to specific vector representations of particular text tokens (e.g. a words or subwords in a sentence). Since attention mechanism processes each vector regardless of its position in the input sequence, each of these representations is usually combined with a positional encoding vector that captures the natural order for each token in the sentence. After that follows a stack of multiple attention layers, the *Transformer encoder layers*. A single Transformer encoder layer receives a fixed set of input vectors $x = [x_1, \ldots, x_n]$ and outputs the same number of vectors $y = [y_1, \ldots, y_n]$ with the same dimensionality. Usually this dimension is set to $d_{\text{model}} = 512$. Due to their uniform format of inputs and outputs, encoder layers can be stacked in an arbitrary sequence to increase model complexity. Thereby, an encoder output serves as the input for the subsequent encoder layer. The final outputs y_j of the encoder stack can then be used as new token representations for task-specific neural network architectures, e.g., for text generation, sequence tagging, or text classification.

Since the Transformer encoder layers have become one of the fundamental building blocks of today's NLP architectures, we will now formally describe them in more detail.

The typical structure of a Transformer encoder layer consists of two main sublayers: a multihead self-attention layer and a position-wise feedforward neural network. The full pipeline is depicted in Fig. 6.4. We will now describe each of the sublayers in more detail.

6.7.1 Self-attention

In the self-attention layer, an attention mechanism is used to generate new vector representations $z = [z_1, \ldots, z_n]$ for a given sequence of input vectors $x = [x_1, \ldots, x_n]$ with $z_i, x_i \in \mathbb{R}^{d_{\text{model}}}$. Since attention refers to words from the same sequence as the word for which we want to generate the new representation, the attention mechanism utilized here is often called *self-attention* [30].

In the self-attention layer, we calculate context vectors c_j from the input sequence x exactly as described in Sect. 6.5. However, in self-attention, the three different "roles" of query, key, and value do not come from different places, but are all calculated from the input x_i which corresponds to the output of the previous encoder layer. Here, each input vector x_i is considered to represent all three different "roles" of query, key, and value simultaneously: $x_i = q_i = k_i = v_i$. Prior to calculating the attention output based on q_i, k_i and v_i, it has been proven to be beneficial to linearly project these vectors into smaller dimensionalities depending on their role. This is done by multiplying the vectors by three projection matrices W^Q, W^K, W^V that will generate query, key, and value vectors of smaller dimensionalities, e.g., $d_K = 64$, $d_V = 64$ with $q_i^\star, k_i^\star \in \mathbb{R}^{d_K}$ and $v_i^\star \in \mathbb{R}^{d_V}$:

$$q_i^\star = W^Q q_i, \quad k_i^\star = W^K k_i, \quad v_i^\star = W^V v_i \tag{6.22}$$

where W^Q, W^K are of size $d_K \times d_{\text{model}}$ and W^V of size $d_V \times d_{\text{model}}$.

It is these linear transformations that define the three roles which a token plays in the attention processing and in the following algorithmic steps. Note that per definition, the attention mechanism itself has no trainable parameters. Instead, the actual training is done by learning the vector representations of queries, keys, and values which requires to learn the elements of weight matrices W^Q, W^K, and W^V.

To create a context vector for position j, a weighted average of all values $v_i^\star$ is formed with respect to weights α_{ji} which express the similarity between input tokens x_j and x_i, or, more exactly between the query $q_j^\star$ for position j and the key $k_i^\star$ for position i (see Sect. 6.5).

$$c_j = \sum_{i=1}^{n} \alpha_{ji} v_i^\star \tag{6.23}$$

Note that by using different matrices for queries W^Q and keys W^K, an asymmetric similarity concept is used. This means that the similarity of output position j to input position i is not equal to the similarity of output position i to output position j. This can be justified in many cases. For example, a noun might be more relevant for encoding a preposition than the other way round. However, it does not necessarily lead to a maximum or even sufficiently high similarity of a token to itself, as would be the case with a symmetric similarity definition. If symmetric case was preferable, it could be reached by using equal matrices $W^K = W^Q$.

By attending both left and right positions in the input sequence, the process of encoding a context vector c_j learns which input positions are highly relevant for this task, thus making the resulting vector c_j aware for bidirectional context. For NLP, it is assumed that using self-attention, textual dependencies can be learned easily, e.g., coreferences. As an example, when encoding the input token "she" in the sentence "Janine told Paul that she liked his presentation", the attention mechanism could learn that the pronoun "she" refers to the person "Janine" and therefore consider this position as highly relevant.

Since we can expect many of these implicit dependencies to appear in text, multiple self-attention mechanisms are commonly applied in parallel. Multiple applications of self-attention mechanisms are referred to as *attention heads* where H is the number of attention heads. Multiple heads can be regarded as multiple views of relevance of individual tokens.

The calculation of the context vectors c_j so far can be seen as the result of one attention head. Different attention heads will perform the same attention operation based on the same input sequence x. However, each attention head $h, h \in [1, \ldots, H]$ has its own set of query, key, and value matrices W_h^Q, W_h^K, and W_h^V which will be learned during the training process. As a result, each head generates its own set of context vectors head$_h = [c_{h1}, \ldots, c_{hn}]$ that may learn different dependencies in a sentence. Here, c_{hj} denotes the context vector for position j and head h.

Finally, the output vectors z_j are generated by transforming all context vectors from all heads back to the larger dimension of d_{model}. This transformation is done by a $d_{\text{model}} \times hd_V$ matrix W^O. More exactly, this matrix can be seen as a horizontal concatenation of head-specific $d_{\text{model}} \times d_V$ matrices W_h^O. If multiplied by vertically stacked context vectors $c_{1j}, \ldots, c_{Hj}$, the transformed output vector at position j is received over all heads:

$$z_j = \begin{bmatrix} W_1^O & \cdots & W_H^O \end{bmatrix} \begin{bmatrix} c_{1j} \\ \vdots \\ c_{Hj} \end{bmatrix} = \sum_{h=1}^{H} W_h^O c_{hj} \tag{6.24}$$

Together with the head-specific computation of context vectors:

$$c_{hj} = \sum_{i=1}^{N} \alpha_{hji} v_{hi}^{\star} = \sum_{i=1}^{N} \alpha_{hji} W_h^V x_i \tag{6.25}$$

we receive:

$$z_j = \sum_{h=1}^{H} W_h^O c_{hj} = \sum_{h=1}^{H} W_h^O \sum_{i=1}^{N} \alpha_{hji} W_h^V x_i = \sum_{h=1}^{H} W_h^O W_h^V \sum_{i=1}^{N} \alpha_{hji} x_i \tag{6.26}$$

We will denote this particular layer of the Transformer *multihead self-attention*:

$$z_j = \text{MultiHeadAtt}(x_j, x) \tag{6.27}$$

The fact that matrices W_h^V and W_h^O appear only in the product $W_h^{OV} = W_h^O W_h^V$ suggests that it may be obsolete to consider these matrices separately. Instead, the product matrix W_h^{OV} would be learned by fitting to data. The decision which alternative is more appropriate depends on the dimension d_V of the value vector and its ratio to the dimension d_{model} of the model vector.

Product matrix W_h^{OV} has d_{model}^2 elements, and thus there are equally many free parameters. Both individual matrices have together $2d_{\text{model}}d_V$ elements. These parameters are partially redundant since taking MW_h^V and $W_h^O M^{-1}$ instead of W_h^V and W_h^O leads to identical results. So the number of effective degrees of freedom is $2d_{\text{model}}d_V - d_V^2$. Obvious consequences are that it is always preferable to seek the product matrix if $d_V \geq d_{\text{model}}$, while it is more economical (and thus better tractable by numerical algorithms) to represent the product by both factor matrices if $d_V < \frac{1}{2}d_{\text{model}}$. The intermediary case is difficult to analyze if there is no possibility to remove the redundancy in matrix elements.

There is also a representational aspect of using relatively small d_V. It amounts to extracting a concise feature set of relatively low dimension d_V from the model vectors with relatively high dimension d_{model}. Attaining this may be itself valuable.

6.7.2   Position-Wise Feedforward Networks

Besides self-attention, each encoder layer uses a feedforward network which transforms the self-attention outputs z_j into new vectors y_j. The network is called *position-wise* since it applies the same transformations for each position j. It consists of two linear transformations and the nonlinear ReLU activation function that processes the output of the first linear transformation. The inner dimension of the hidden layer d_{inner} is usually of higher dimensionality than the model dimension of the input and output d_{model}, e.g., $d_{\text{inner}} = 2048$ with $d_{\text{model}} < d_{\text{inner}}$.

$$y_j = \text{PFF}(z_j) = W_2 F(W_1 z_j + b_1) + b_2 \tag{6.28}$$

where F is the ReLU activation function:

$$F(x) = \max(0, x) \tag{6.29}$$

6.7.3   Residual Connection and Layer Normalization

In the previous subsections, the main functional components of Transformer have been described. Besides them, some architectural features are worth mentioning that have the aim of improving the convergence of numerical optimization algorithms. They are employed within the Transformer encoder layer but can also frequently

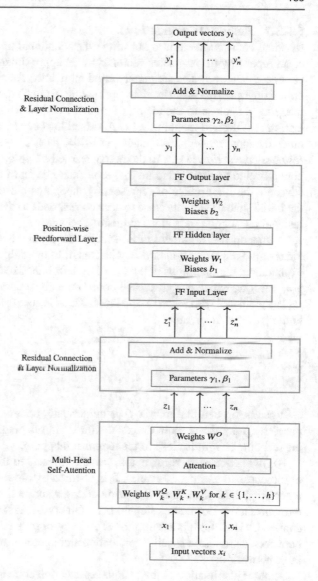

Fig. 6.4 Transformer encoder layer

be encountered outside of Transformer architecture. Commonly used measures include residual connections and layer normalizations. For example, Vaswani et al. [30] use residual connections, followed by layer normalization at two points in the encoder layer: after self-attention and after the position-wise feedforward network (see Fig. 6.4).

6.7.3.1   Residual Connections

Residual connections aim to substitute the functional approximator $y = f(x, p)$ of an input x with parameter vector p by an approximator $y = x + g(x, q)$ with parameter vector q. The frequently cited motivation for this is by the intuition that optimizing $g(x, q)$ is easier than optimizing $f(x, p)$, especially when the sought mapping f is close to the identity function $\text{id}(x) = x$. This means that instead of learning the desired mapping $f(x, p)$ directly, $g(x, q)$ learns the difference of how much the input x needs to change: $g(x, q) = f(x, p) - x$. By adding the so-called *residual connection* [11] x to the newly learned mapping g, the identity function is now modeled explicitly and no longer needs to be learned by f. Some researchers advocate the importance of representing mappings close to identity. In particular, the initial state of weights close to zero corresponds to the zero function, while with residual connection, it is close to identity.

To support this argument more exactly, the gradients can be observed. The gradient of the objective function E with regard to the output y is $\frac{\partial E}{\partial y}$. For easiness of explanation, the mappings will be used in their linearized form, that is $y = Px$ and $y = x + Qx$, which can always be done at the given parameter point.

The gradients with regard to the input, which is necessary for the backpropagation to the previous layer, are

$$\frac{\partial E}{\partial y}\frac{\partial y}{\partial x} = \frac{\partial E}{\partial y}P \tag{6.30}$$

and

$$\frac{\partial E}{\partial y}\frac{\partial y}{\partial x} = \frac{\partial E}{\partial y}(I + Q) \tag{6.31}$$

It can be argued that matrix Q is numerically better conditioned than P, which possesses a dominant main diagonal with small off-diagonal elements—as long as the mapping sought is close to the identical mapping.

For dynamical, i.e., recurrent, systems as presented in Sect. 2.3, the assumption of closeness to identity is natural. It is justified by the fact that the state-space representation 2.73 of the transformation during a single time step mostly implies that the state change during this step is small. Otherwise, fast subprocesses would not be captured.[1] How far this assumption is also appropriate for static, non-recurrent deep networks, depends on specific application architectures and good reasons in favor of large network depth.

Another justification of the residual connection concept is the vanishing gradient phenomenon in deep (i.e., having many stacked hidden layers) networks. The h-th layer can be represented as a mapping

$$y = Ix + F_h(x) \tag{6.32}$$

with I being the identity matrix, x the layer input, and y the layer output. Assuming linearity (e.g., if the mapping F is linearly approximated around its argument x), 6.32 can be written in the operator notation

$$y = Ix + F_h x = (I + F_h)x \tag{6.33}$$

---

[1] This is the consequence of the Nyquist theorem known in the control theory.

where Fx corresponds to the function representation $F(x)$. In this notation, nested functions $F(G(x))$ are written as a product FGx.

The stack of such layers for $h = 1, \ldots, H$ is represented by the product

$$y = \left(\prod_{h=1}^{H} (I + F_h) \right) x \tag{6.34}$$

The operator in parentheses can be expanded to the form

$$I + \sum_{h \le H} F_h + \sum_{i < j \le H} F_j F_i + \cdots + \prod_{h=H,\ldots,1} F_h \tag{6.35}$$

It is a sum of products of H operators F_h, for $h = 1, \ldots, H$, which includes all single operators F_h, all pairs, all triples, etc. A conventional, non-residual network would correspond to the last term $\prod_{h=H,\ldots,1} F_h$. Here, the reverse index order results from the nesting order—operators with lower indices are nested within (and therefore executed before) operators with higher indices.

In terms of neural network architecture, the decomposition 6.35 can be thought of as a set of networks with a common input and an output formed by the sum over all networks of the set. The networks would be, in particular

- an identity single-layer network (with a unit matrix of weights);
- H single-layer networks (with no hidden layer);
- $\frac{H(H-1)}{2}$ two-layer networks (with one hidden layer);
- corresponding numbers of multilayer networks of depths 3 to $H - 1$;
- one network with H layers (i.e., $H - 1$ hidden layers).

(The artificial distinction between the different networks of the mentioned set is, of course, only hypothetical to illustrate the properties of a network with residual layers.)

This representation makes clear how the vanishing gradient problem is alleviated. The size of gradient (i.e., derivative of the network output with regard to the network input) tends to decrease with chaining the layers, resulting in negligibly small values in deep networks. By contrast, the residual network in the decomposition 6.35 also contains the term $\sum_h F_h$. The network output is thus partially determined by the sum of outputs of individual layers. The gradient of this term is as large as that of a single layer. This prevents the gradient sum from vanishing.

With residual networks, tractable networks with many (even hundreds) hidden layers have been successfully trained. However, to assess the merit of this architectural idea, it has to be kept in mind that a network of the form 6.35 is obviously a mixture of

- a shallow network with multiple branches (the term $\sum_h F_h$);
- a deep network (the term $\prod_{h=H,\ldots,1} F_h$); and
- all intermediary network depths.

It is hard to assess whether the success of such networks is to be attributed to the shallow or to the deep component. The gradient of the shallow one is, under configurations with vanishing gradient, clearly dominating while that of the deep one would still vanish. Consequently, the original "formally deep" network 6.34 which is equivalent to the expansion 6.35 behaves approximately like a sum of parallel single-layer networks (corresponding to $\sum_h F_h$). This expectation is based on the fact that vanishing gradient indicates a small influence of the involved parameters on the mapping output, while large gradients testify a strong influence. This may suggest that it is not the depth of the network but its special structure that is responsible for the success.

In summary, numerical advantage (supported by computing experience) can be attained.

Natural restrictions in using the residual connection concept consist in

- the requirement that x and y have the same dimension; and
- the expectation that the mapping materialized by a single layer will remain close to identity mapping even for the optimum fit to data.

The latter restriction is alleviated with growing depth of the network. The question remains, whether deep subnetworks with forced identical layer dimensions and submappings close to identity are generally superior to shallow subnetworks without such restrictions.

In the Transformer encoder layer, residual connections are used to substitute the multihead self-attention function

$$z_j = \text{MultiHeadAtt}\left(x_j, x\right) \tag{6.36}$$

with:

$$z_j = \text{MultiHeadAtt}\left(x_j, x\right) + x_j \tag{6.37}$$

and the position-wise feedforward layer function:

$$y_j = \text{PFF}\left(z_j\right) \tag{6.38}$$

with:

$$y_j = \text{PFF}\left(z_j\right) + z_j \tag{6.39}$$

6.7.3.2 Normalization Concept and Its Motivation

Input variables to an approximator may have different scaling. For a linear model such as linear regression scaling is accounted for by pseudoinversion of matrix $X'X$, used in the regression formula 4.9. However, scaling differences of several orders of magnitude may result in a bad numerical conditioning of the inverse, since the matrix $X'X$ contains elements of very different size.

Typical neural network approximator architectures consist of stacked modules whose input is computed as output of the previous layer. Then, an additional problem arises that non-normalized outputs of one layer may hit problematic regions of nonlinear activation functions such as sigmoid or rectifier. These are motivations for *normalization of input or hidden variables*.

Furthermore, parameters of layers with arbitrarily scaled variables are not sufficiently determined. Even in the case of consecutive linear layers with weight matrices W_i and W_{i+1}, the processing is identical with weights MW_i and $W_{i+1}M^{-1}$, for arbitrary non-singular square matrix M with appropriate dimension, since $W_{i+1}W_i = W_{i+1}M^{-1}MW_i$. Consequently, both operators are equal. This makes clear that layer matrices are unbounded: either MW_i or $W_{i+1}M^{-1}$ may arbitrarily grow by choosing matrix M with a very large or very small norm. Such unbounded parameters may be a serious problem for the numeric optimization.

A possible remedy is normalizing columns of W_i. It makes the representation unique except for orthonormal transformations which do not affect the matrix norm and thus the magnitude of the intermediary representation by W_i. This motivates *normalization of network parameters*.

Both motivations are interconnected—unbounded network parameters would lead to unbounded activations of hidden layers (e.g., in the linear segment of rectifier units) or to saturation (e.g., in the sigmoid activation function or the constant segment of rectifier units). The problem grows with the depth of the network—divergent values can be attenuated through the stack of layers. These reasons make the normalization in deep networks desirable. The latter form of normalization, that of network parameters, is closely related to the regularization concept of Sect. 4.7. In the following, some remarks to the normalization of variables will be made. A widespread normalization principle is a transformation to zero mean and unity variance:

$$\hat{y} = \frac{z - \mu}{\sqrt{v}} \tag{6.40}$$

with m being the mean of variable z and v its variance, computed over the batch training set of size H. Since both the mean and the variance are a priori unknown and additionally varying with layer's parameters (which change during the fitting process), they are estimated from a sample, typically the training set or its part. This normalization method is referred to as batch normalization. Usually, it is supplied an additional scaling and shift parameter. Computing the gradient of the fit measure E (e.g., MSE of the output layer), it would be $\frac{\partial E}{\partial y}$ without normalization. This vector would be propagated backward through the layer's weight matrix. The computation requires only the values of the present training sample, and all sample gradients will ultimately be summed up to deliver the gradient over the training set. With layer output y standardized, this extends to

$$\begin{aligned}\frac{\partial E}{\partial z} &= \frac{\partial E}{\partial y}\left[\left(1 - \frac{\partial m}{\partial z}\right)\frac{1}{\sqrt{v}} - (z - m)\frac{\partial v}{\partial z}\frac{1}{2v\sqrt{v}}\right] \\ &= \frac{\partial E}{\partial y}\frac{1}{\sqrt{v}}\left[\left(1 - \frac{1}{H}\right) - (z - m)^2\frac{1}{Hv}\right]\end{aligned} \tag{6.41}$$

Besides scaling by standard deviation $\sqrt{v}$, the term in outer brackets is smaller than unity. This is why the derivative can be expected to be smaller with batch normalization than without it.

Consistently with this, Santurkar et al. [26] have found upper bounds for the norms of both the gradient and the Hessian matrix of second derivatives with regard to neural network parameters. This indicates that the mapping is made smoother with help of batch normalization. A tendency to smoothing has also been observed on gradient norm development during optimization. Although this may strongly depend on the optimization method used (first or second order, variants of step size control), batch normalization seems to be a helpful tool.

The gradient expression for batch normalization has the property that the gradient for a single sample depends on all training samples. A lot of simplicity of non-normalized gradient computation is lost. This has led to proposing alternative normalization concepts, with less stringent theoretical support, such as layer normalization. It is being used in Transformer architecture and is described in more detail in the next subsection.

6.7.3.3 Layer Normalization in Transformer Architecture

The Transformer encoder uses a simplified concept of normalization. The principle is consistent with (6.40), a transformation to zero mean and unity variance:

$$\hat{x} = \frac{x - \mu}{\sqrt{\sigma^2}} \tag{6.42}$$

with μ being the mean of variable x and σ^2 its variance. The notation adopts to that used for Transformers: $\hat{x}$ and x corresponds to y and z in (6.40), respectively.

In the Transformer, normalization is applied after the residual connections of the multihead self-attention layer and the position-wise feedforward layer. In both cases, the variables to be normalized refer to the sum of input and output vectors of the multihead self-attention layer:

$$z_j^* = \text{LayerNorm}\left(z_j + x_j\right) = \text{LayerNorm}\left(\text{MultiHeadAtt}\left(x_j, x\right) + x_j\right) \tag{6.43}$$

and the position-wise feedforward layer:

$$y_j^* = \text{LayerNorm}\left(y_j + z_j^*\right) = \text{LayerNorm}(\text{PFF}\left(z_j^*\right) + z_j^*) \tag{6.44}$$

In contrast to *batch normalization* [15], where mean and variance are calculated over the training data or small groups of samples (so-called *mini-batches*), *layer normalization* [1] calculates these statistics over the feature dimension. The difference between batch and layer normalization is visualized in Fig. 6.5.

The choice of layer normalization in the Transformer architecture can be traced back to its established use in the field of NLP, where empiric studies have shown their superiority to batch normalization methods, mainly when used with recurrent approaches, which were the main approaches used before the rise of the Transformer architectures. One reason for using layer normalization in NLP is the fact that, contrary to image processing, sentences of different sequence lengths need to be processed. If using batch normalization, the optimization constants (the number of values over which mean and variance need to be calculated) would change over

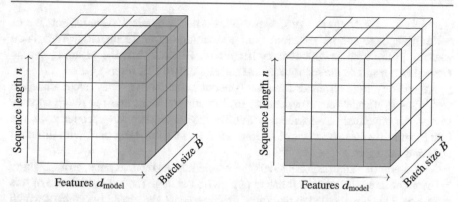

Fig. 6.5 Difference of batch normalization (*left*) and layer normalization (*right*). The highlighted areas show the values that are used to calculate the normalization statistics μ and σ^2

different batches. This means that "[…] statistics of NLP data across the batch dimension exhibit large fluctuations throughout training. This results in instability, if batch normalization is naively implemented." [28]. However, the question of whether layer normalization is the best normalization in the Transformer still requires mathematical discussion.

For any vector $v \in \mathbb{R}^{d_{\text{model}}}$, layer normalization is defined as follows:

$$\text{LayerNorm}\,(v) = \gamma \frac{v - \mu}{\sqrt{\sigma^2 + \epsilon}} + \beta \tag{6.45}$$

where ϵ is a constant for numerical stability and the mean and variance are determined by:

$$\mu = \frac{1}{d_{\text{model}}} \sum_{i=1}^{d_{\text{model}}} v_i \tag{6.46}$$

and

$$\sigma^2 = \frac{1}{d_{\text{model}}} \sum_{i=1}^{d_{\text{model}}} (v_i - \mu)^2 \tag{6.47}$$

Simply normalizing the vector to zero mean and unity variance could constrain the inputs just to a specific subset of the activation function (e.g., the linear part of a nonlinear function). Therefore, two learnable parameters γ and β are introduced, which ensure that the identity function can be learned by scaling and shifting the normalized value.

6.8 Section Summary

Natural language is a sequence of consecutive tokens (e.g., words) and can thus be viewed as a dynamical system. The meaning of the sequence is encoded in this flow. A model of natural language has the sequence of tokens as its input and the meaning (semantics) as its output. Consistently with other dynamical systems, models of natural language can be recurrent (i.e., containing output feedback), which is an analogy to the transfer functions or IIR filters. Alternatively, the meaning can be viewed as a result of the input sequence (an analogy to FIR filters).

An analogy to the transfer function concept is the RNNs. They accumulate the semantics in a state vector depending on the current input and the previous state vector. An amended development with advantageous computing properties is the LSTM splitting the state vector to a long-term and a short-term part, with different updating strategies.

The alternative approach considers the whole text segment at once, as an analogy to the mentioned FIR model. It has to cope with the huge (and variable) size of this sequence. This obstacle led to particular developments, the most important of which is referred to as attention mechanism. Mathematically, the concept is based on vector similarity.

The attention mechanism is used in the successful architecture of Transformers. The attention mechanism selects tokens with the high context relevance for a particular token, reducing the size of the context information considered. The semantic vector composed of the relevant context tokens is processed by a nonlinear feedforward network. Usually, a sequence of multiple Transformer layers is used.

Residual connections and normalizations are auxiliary concepts used to improve the numeric optimization properties of the model.

6.9 Comprehension Check

1. Which are the two alternatives for modeling natural language as a dynamical system?
2. Which of these alternatives is adopted by the LSTM?
3. Which of these alternatives is followed by the Transformer architecture?
4. Which are the typical measures for semantic similarity?
5. What is the main goal of the attention mechanism?
6. By which means is the similarity measure in Transformer architecture made asymmetric?

References

1. Ba JL, Kiros JR, Hinton GE (2016) Layer normalization. https://doi.org/10.48550/arXiv.1607.06450
2. Bahdanau D, Cho K, Bengio Y (2016) Neural machine translation by jointly learning to align and translate. ICLR. http://arxiv.org/abs/1409.0473

3. Baroni M, Lenci A (2010) Distributional memory: a general framework for corpus-based semantics. Comput Linguist 36(4):673–721. https://doi.org/10.1162/colispsasps00016

4. Berard A, Pietquin O, Servan C, Besacier L (2016) Listen and translate: a proof of concept for end-to-end speech-to-text translation. https://doi.org/10.48550/arXiv.1612.01744

5. Bojanowski P, Grave E, Joulin A, Mikolov T (2017) Enriching word vectors with subword information. Trans Assoc Comput Linguis 5:135–146. https://doi.org/10.1162/taclspsasps00051

6. Celikyilmaz A, Clark E, Gao J (2021) evaluation of text generation: a survey. http://arxiv.org/abs/2006.14799

7. Cho K, van Merriënboer B, Gulcehre C, Bahdanau D, Bougares F, Schwenk H, Bengio Y (2014) Learning Phrase Representations using RNN encoder–decoder for statistical machine translation. In: Proceedings of the 2014 conference on empirical methods in natural language processing (EMNLP), Association for Computational Linguistics, Doha, Qatar, pp 1724–1734. https://doi.org/10.3115/v1/D14-1179

8. Devlin J, Chang MW, Lee K, Toutanova K (2019) BERT: Pre-training of deep bidirectional transformers for language understanding. In: Proceedings of the 2019 conference of the North American chapter of the Association for Computational Linguistics: Human Language Technologies, volume 1 (Long and Short Papers), Association for Computational Linguistics, Minneapolis, Minnesota, pp 4171–4186. https://doi.org/10.18653/v1/N19-1423

9. Gers FA, Schmidhuber J, Cummins F (2000) Learning to forget: continual prediction with LSTM. Neural Comput 12(10):2451–2471. https://doi.org/10.1162/089976600300015015

10. Graves A (2014) Generating sequences with recurrent neural networks. https://doi.org/10.48550/arXiv.1308.0850

11. He K, Zhang X, Ren S, Sun J (2016) Deep residual learning for image recognition. In: 2016 IEEE conference on computer vision and pattern recognition (CVPR), pp 770–778. https://doi.org/10.1109/CVPR.2016.90

12. Hirschberg J, Manning CD (2015) Advances in natural language processing. Science 349(6245):261–266. https://doi.org/10.1126/science.aaa8685

13. Hochreiter S, Schmidhuber J (1997) Long short-term memory. Neural Comput 9(8):1735–1780. https://doi.org/10.1162/neco.1997.9.8.1735

14. Hochreiter S, Bengio Y, Frasconi P, Schmidhuber J (2001) Gradient Flow in Recurrent Nets: The Difficulty of Learning Long-Term Dependencies. In: Kolen JF, Kremer SC (eds) A Field Guide to Dynamical Recurrent Networks, Wiley-IEEE Press, pp 237–243. https://doi.org/10.1109/9780470544037

15. Ioffe S, Szegedy C (2015) Batch Normalization: accelerating deep network training by reducing internal covariate shift. https://doi.org/10.48550/arXiv.1502.03167

16. Jing K, Xu J (2019) A survey on neural network language models. http://arxiv.org/abs/1906.03591

17. Kim Y, Denton C, Hoang L, Rush AM (2017) Structured attention networks. ICLR. https://arxiv.org/abs/1702.00887v3

18. Klein G, Kim Y, Deng Y, Senellart J, Rush A (2017) OpenNMT: open-source toolkit for neural machine translation. In: Proceedings of ACL 2017, system demonstrations, Association for Computational Linguistics, Vancouver, Canada, pp 67–72. https://doi.org/10.18653/v1/P17-4012

19. Klopfenstein LC, Delpriori S, Malatini S, Bogliolo A (2017) The rise of bots: a survey of conversational interfaces, patterns, and paradigms. In: Proceedings of the 2017 conference on designing interactive systems, Association for Computing Machinery, New York, NY, USA, DIS '17, pp 555–565. https://doi.org/10.1145/3064663.3064672

20. Manning CD, Raghavan P, Schütze H (2008) Introduction to information retrieval, 1st edn. Cambridge University Press. https://doi.org/10.1017/CBO9780511809071

21. Niklaus C, Cetto M, Freitas A, Handschuh S (2018) A survey on open information extraction. In: Proceedings of the 27th international conference on computational linguistics, Association for Computational Linguistics, Santa Fe, New Mexico, USA, pp 3866–3878. https://www.aclweb.org/anthology/C18-1326

22. Pennington J, Socher R, Manning C (2014) Global vectors for word representation. In: Proceedings of the 2014 conference on empirical methods in natural language processing (EMNLP), Association for Computational Linguistics, Doha, Qatar, pp 1532–1543. https://doi.org/10.3115/v1/D14-1162

23. Peters ME, Neumann M, Iyyer M, Gardner M, Clark C, Lee K, Zettlemoyer L (2018) Deep contextualized word representations. In: Proceedings of the 2018 conference of the North American chapter of the Association for Computational Linguistics: Human Language Technologies, volume 1 (long papers), Association for Computational Linguistics, New Orleans, Louisiana, pp 2227–2237. https://doi.org/10.18653/v1/N18-1202

24. Prakash A, Hasan SA, Lee K, Datla V, Qadir A, Liu J, Farri O (2016) Neural paraphrase generation with stacked residual LSTM networks. In: Proceedings of COLING 2016, the 26th international conference on computational linguistics: technical papers, The COLING 2016 Organizing Committee, Osaka, Japan, pp 2923–2934. https://aclanthology.org/C16-1275

25. Salton G, Wong A, Yang CS (1975) A vector space model for automatic indexing. Commun ACM 18(11):613–620. https://doi.org/10.1145/361219.361220

26. Santurkar S, Tsipras D, Ilyas A, Mądry A (2018) How does batch normalization help optimization? In: Proceedings of the 32nd international conference on neural information processing systems. Curran Associates Inc., Red Hook, NY, USA, NIPS'18, pp 2488–2498

27. Schuster M, Paliwal K (1997) Bidirectional recurrent neural networks. IEEE Trans Signal Process 45(11):2673–2681. https://doi.org/10.1109/78.650093

28. Shen S, Yao Z, Gholami A, Mahoney M, Keutzer K (2020) PowerNorm: Rethinking batch normalization in transformers. In: III HD, Singh A (eds) Proceedings of the 37th international conference on machine learning, PMLR, Proceedings of machine learning research, vol 119, pp 8741–8751.https://proceedings.mlr.press/v119/shen20e.html

29. Soltau H, Liao H, Sak H (2017) neural speech recognizer: acoustic-to-word LSTM model for large vocabulary speech recognition. In: Interspeech 2017, ISCA, pp 3707–3711. https://doi.org/10.21437/Interspeech.2017-1566

30. Vaswani A, Shazeer N, Parmar N, Uszkoreit J, Jones L, Gomez AN, Kaiser Ł, Polosukhin I (2017) Attention is all you need. In: Proceedings of the 31st International conference on neural information processing systems. Curran Associates Inc., Red Hook, NY, USA, NIPS'17, pp 6000–6010

31. Wang S, Zhou W, Jiang C (2020) A survey of word embeddings based on deep learning. Computing 102(3):717–740. https://doi.org/10.1007/s00607-019-00768-7

32. Wang ZQ, Tashev I (2017) Learning utterance-level representations for speech emotion and age/gender recognition using deep neural networks. In: 2017 IEEE international conference on acoustics, speech and signal processing (ICASSP), IEEE Press, New Orleans, LA, USA, pp 5150–5154. https://doi.org/10.1109/ICASSP.2017.7953138

33. Wu Y, Schuster M, Chen Z, Le QV, Norouzi M, Macherey W, Krikun M, Cao Y, Gao Q, Macherey K, Klingner J, Shah A, Johnson M, Liu X, Kaiser Ł, Gouws S, Kato Y, Kudo T, Kazawa H, Stevens K, Kurian G, Patil N, Wang W, Young C, Smith J, Riesa J, Rudnick A, Vinyals O, Corrado G, Hughes M, Dean J (2016) Google's neural machine translation system: bridging the gap between human and machine translation. https://doi.org/10.48550/arXiv.1609.08144

34. Yogatama D, Dyer C, Ling W, Blunsom P (2017) Generative and discriminative text classification with recurrent neural networks. https://doi.org/10.48550/arXiv.1703.01898

35. Zhao M, Dufter P, Yaghoobzadeh Y, Schütze H (2020) Quantifying the contextualization of word representations with semantic class probing. In: Findings of the Association for Computational Linguistics: EMNLP 2020, Association for Computational Linguistics, Online, pp 1219–1234. https://doi.org/10.18653/v1/2020.findings-emnlp.109

Specific Problems of Computer Vision

7

Computer Vision (CV) covers the whole variety of collecting and interpreting information from optical signals of any type. A classical review is for example Wechsler [9]. Its currently largest volume of applications counts to the field of image processing: evaluating the scene represented by a digital image. A digital image is usually represented by a two-dimensional, rectangular matrix of pixels, each of which is assigned a value, or vector of values, corresponding the gray shade or a color vector.

Interpreting the content of a digital image embraces a variety of partial tasks such as

(a) extracting and recognizing edges delimiting interesting parts of the image;
(b) normalizing the image of its parts in regard to contrast and brightness, to account for recognizing the same objects in different light conditions;
(c) segmenting the image to distinct parts probably corresponding to distinct real-world objects.
(d) accounting for shift, rotation, scaling, and further invariances to recognize the same objects modified (e.g., shifted and rotated) views;
(e) recognizing individual objects and their attributes.

From these tasks, (a), (b), and (d) are mathematical operations, while (c) and (e) consist in nominal labeling. Step (e) can frequently be viewed as a classification task as long as the label set is predefined.

For such tasks such as edge recognition, mathematical operators on the environment of a given pixel (i.e., pixels close to the given one) have been found. An example is the well-known *Laplacian* edge detector. The principle of such operators is explained in Sect. 3.6. Also, the normalization of contrast and brightness can be formulated in a mathematical way.

Invariances such as shift and rotation are discussed in more detail in Sect. 7.2.

© The Author(s), under exclusive license to Springer Nature Switzerland AG 2023
T. Hrycej et al., *Mathematical Foundations of Data Science*, Texts in Computer Science,
https://doi.org/10.1007/978-3-031-19074-2_7

7.1 Sequence of Convolutional Operators

Convolutional layers are based on two ideas:

1. connections between layers restricted in a particular way; and
2. shared parameters.

The first one is that for some tasks, it is useful to restrain the connections between individual units of consecutive layers to those of a small neighborhood. More exactly, the units α of a layer A are connected to the units of the predecessor layer B that correspond to the neighborhood of α. This results in the requirement that both layers have to possess a common structure, to be able to determine which unit of layer B corresponds to a unit of layer A. This common structure has also to contain the same neighborhood concept.

A 2D image consists of a matrix of pixels. Most straightforward definitions of a neighborhood are defined in terms of difference between the indices along both matrix dimensions. The pixel $a_{i,j}$ can have eight neighbors $a_{i-1,j-1}, a_{i-1,j}, a_{i-1,j+1}, a_{i,j-1}, a_{i,j+1}, a_{i+1,j-1}, a_{i+1,j}, a_{i+1,j+1}$, all having the maximum difference of an element index of one. This is shown in Fig. 7.1.

A unit of the kth layer with indices i, j would then be connected to only the units of the $(k-1)$th layer corresponding to the highlighted region of Fig. 7.1 (including the red central unit).

The idea of shared parameters consists in connection weights common to all neighborhoods in the pixel matrix. In this example, the set of weights connecting the unit of the kth layer with the nine units of the $(k-1)$th layer is shared for all index pairs (i, j). This has been formally presented in Sect. 3.6.

It is obvious that such layer organization is a direct implementation of a local operator. With appropriate weights, well-known operators such as the Laplace edge

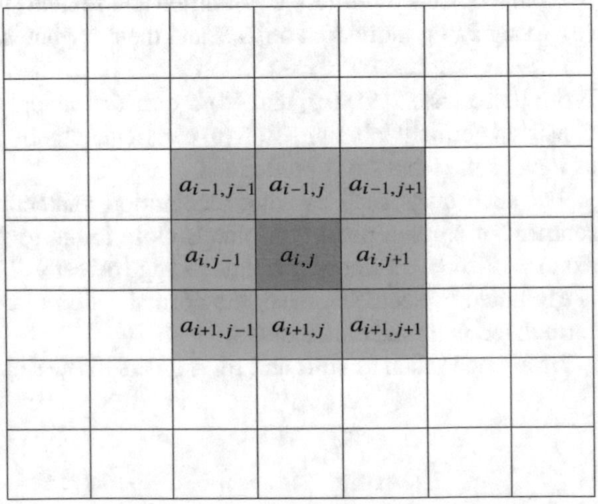

Fig. 7.1 2D neighborhood in a pixel matrix

detector can be implemented. It is a straightforward idea that local operators of this local type can be learned by parameter fitting.

The definition of a particular environment together with the corresponding weights corresponds to the general concept of *kernel*. With *Convolutional Neural Networks* (CNNs), it is frequently called *filter*. This term will be used in the following text.

More complex, nonlinear operators can be implemented by a sequence of such convolutional layers.

7.1.1   Convolutional Layer

As defined in 3.29 for an example neighborhood of 3×3, a convolutional layer is defined with the help of an operator implemented as a filter with randomly initialized weight parameters. This operator is repeated with the same parameters for all neighborhoods of each pixel from the preceding layer's output. For example, one such operators parameters might be fitted to be the Laplace edge detector from 3.31. The overlap of the filter is chosen by the user.

The term *convolution* requires that the filter is applied to every pixel so that the size of the layer input is identical with that of layer output. Consequently, adjacent neighborhoods are shifted by one pixel and overlap completely except for one row or column.

However, there are implementations where the shift (called *stride*) is larger than one, up to non-overlapping neighborhoods, with various justifications.

Convolutional layers can be stacked in a sequence, each layer implementing a certain image transformation. Another possibility is to process multiple filters in parallel and concatenate them into a layer of extended size, in which multiple transformed images are placed next to each other. Layer parts corresponding to individual filters are called *feature maps*.

Let's consider the data set *MNIST* [5] which contains 60000 images of handwritten digits labeled into their respective classes 0 to 9. Each of them is of the same quadratic size of 28×28 pixels. The images are grayscale, so the matrix representations are in 2D with values between 0 (black) and 1 (white). Figure 7.2 shows one example image, labeled as the digit *4*.

The classification model is built with one convolutional layer with eight filters and each filter is of size 5×5. The filters' parameters are initialized randomly from a uniform distribution centered around zero. So, they can also have negative values.

Fig. 7.2 Example image of size 28×28 from the MNIST data set depicting a *4*

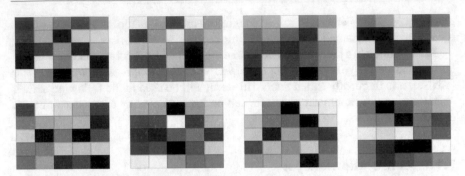

Fig. 7.3 Eight randomly initialized convolutional filters of size 5 × 5 before training

Fig. 7.4 Eight randomly initialized convolutional filters from Fig. 7.3 applied on the example image Fig. 7.2

The filters are visualized in Fig. 7.3. As a result of random initialization of filter weights, there are no obvious patterns.

When applying each filter with a stride of one—i.e., with overlapping neighborhoods—the resulting eight intermediate feature maps can be seen in Fig. 7.4. Activations are varying from black (low activation values) to white (high values). Randomly initialized filters do not extract obviously useful features from the image.

This becomes even clearer after applying the ReLU activation function in Fig. 7.5. There is no clear image processing feature except for the top right where the filter is coincidentally reacting to left borders.

After fitting the model to the data set by adjusting the filter values by minimizing the configured loss function, the filters change into Fig. 7.6.

The filters show more structure than the random ones but are still not easy to interpret—the patterns that the filters react to are not immediately obvious. However, applying them on the example image results in activations (Fig. 7.7). Passing the activations through ReLU leads to Fig. 7.8. Both figures show that the filters extract specific features. They detect various border orientations:

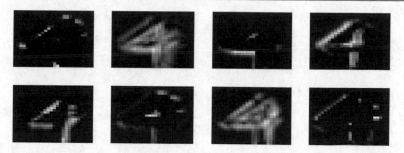

Fig. 7.5 Eight randomly initialized convolutional filter applied on the example image from Fig. 7.2 activated with a ReLU

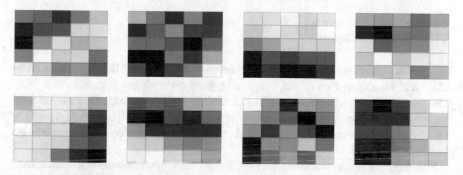

Fig. 7.6 Eight convolutional filters of size 5×5 after training

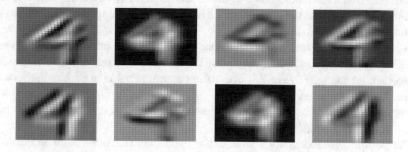

Fig. 7.7 Eight fitted convolutional filters applied on the example image from Fig. 7.2

- Filters 1 and 8: left borders
- Filter 5: right borders
- Filter 3: bottom borders
- Filter 6: top borders.

The remaining filters may react to other image content like round shapes from other digits.

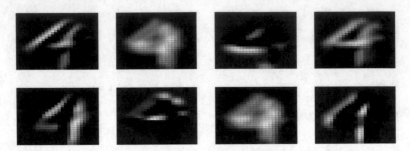

Fig. 7.8 Eight fitted convolutional filters applied on the example image from Fig. 7.2 activated with a ReLU

7.1.2   Pooling Layers

The hierarchical character of image interpretation suggests that the processing starts with local characteristics of the image and goes on with more global ones. This principle can be materialized through successively widening the receptive fields of operators. A widespread implementation of this idea is through including a *pooling* layer after the convolution operator. It decreases the input size of the subsequent convolutional layer.

Similarly to the convolutional operator, the pooling operator works on a predefined neighborhood size. In contrast to the former, where a stride of one (which is the only variant consistent with the concept of convolution) and strong overlapping is usual, pooling operators mostly work with no or reduced overlapping, to reach the goal of reducing the size of processed data.

A common pooling operator is *maximum pooling* which takes the maximum value of each neighborhood, effectively reducing the size of the output matrix. This process is shown as an example in Fig. 7.9 where the maximum value in each neighborhood of size 2×2 is returned.

In general, there is no restriction on the size or stride. It is not common to pool over a non-square neighborhood.

The pooling operator is not limited to the maximum function. Some applications of pooling layers compute the average value for each neighborhood.

The minimum function could also be applied. However, with a widespread rectified linear activation function (ReLU, see Fig. 3.2) whose minimum is by definition zero, minimization would result in a propagation of zeros throughout the layers.

7.1.3   Implementations of Convolutional Neural Networks

In recent years, large labeled image data sets have been collected. They have been used for extensive comparative studies of CV approaches. Some comparisons have taken place in the form of contests (e.g., the ImageNet Large Scale Visual Recognition Challenge (ILSVRC) [6]). The importance of convolutional layers in these

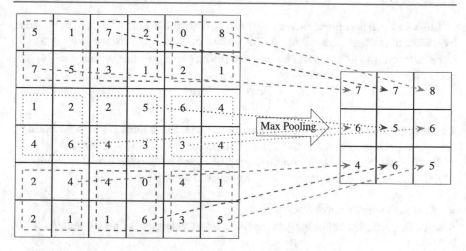

Fig. 7.9 Example of a maximum pooling operator of size 2 × 2 and a stride of 2

comparisons is perpetually growing. With *AlexNet* [4] and subsequently *VGG19* [8], they seem to be superior to other image classification approaches presented so far.

There is currently a trend to make the networks deeper, i.e., using more stacked hidden convolutional layers. The problem of vanishing gradient when stacking additional nonlinear layers halted the development for a short time. The introduction of *residual connections* (see Sect. 6.7.3.1) alleviated this problem with so-called Residual Neural Networks *(ResNet)* [2] which are CNNs with 50, 101, 152, or even 1000 [3] convolutional layers. In many cases, they became superior in comparative studies.

7.2 Handling Invariances

An essential problem in CV is the treatment of invariances. Human and animal vision is characterized by reliably recognizing objects in various conditions. Among the most important invariances, there are

- positions of the object in the visual field or image;
- rotations of the object in two or three dimensions;
- scaling of the object, depending on the individual size and the distance from the observer;
- differences in brightness depending on the current illumination conditions; and
- for many objects, coloring can be varying.

From a mathematical point of view, many invariances correspond to relatively simple operations.

Shift and rotation invariance can be reached with the help of a few parameters by geometrical transformations. A 2D vector of pixel positions $x = \begin{bmatrix} x_1 & x_2 \end{bmatrix}$ is transformed through a rotation by angle ω into a pixel position vector $y = \begin{bmatrix} y_1 & y_2 \end{bmatrix}$ according to

$$y = \begin{bmatrix} \cos\omega & -\sin\omega \\ \sin\omega & \cos\omega \end{bmatrix} x \qquad (7.1)$$

The task is to find the angle ω for which the object is rotated back to its basic or canonical position. For a 2D shift, two obvious parameters are sufficient.

In this way, the invariances might be implemented by varying a moderate number of parameters during the recognition. However, this may be

- too slow for real-time operation; and
- difficult to implement as long as the position of the object to be recognized is not known.

It has also to be noted that this does not seem to be the mechanism by which the vision system of mammals materializes recognition of rotated and shifted objects. Their rotation invariance is not perfect: It is easier to recognize moderately slanted objects than those rotated upside down.

To a small extent, local invariances can be implemented by convolutional layers. In the case of a position shift by one pixel, the convolution operator would consist in having the weight of one to the neighbor pixel in the previous pixel and zero weights otherwise. Slightly more complex but straightforward would be the implementation of a rotation by a fixed angle around a fixed center of rotation. The problem of this approach is the necessity of having a large variety of such convolution operators for all possible shifts, rotation angles, etc.

A fundamental mathematical approach is the *tangent propagation* [7]. It consists in extending the error function by a term penalizing a small change along a desired invariance path. In this way, those mappings are favored which are insensitive to the changes for which they are to be invariant. For example, if position invariance along the horizontal axis is desired, the penalty term would penalize the increase of error function if a pixel value is substituted by its horizontal neighbor value in the input image.

The name tangent propagation is motivated by the idea of describing the invariance by a defined artificial parameter along which the invariance should hold. For the horizontal position invariance, this would be the position x. *Tangent* describes the direction of a small change (i.e., the tangent to the pixel value function if the position changes by a small amount from the current position in the input image).

More formally, every pixel (i, j) is assigned a function $p_{ij}(\delta x)$ of position shift δx along the x axis. For a zero shift, $p_{ij}(0)$ is equal to the actual pixel value in the image. For a small shift of the size of one pixel width, the function $p_{ij}(1)$ equals the value of pixel $(i, j + 1)$. With interpolation, the function can be assigned values for fractional shifts δx. The error function penalty term F may be, for example, the sum of the squares of the partial derivatives

$$\frac{\partial F}{\partial \delta x} = \frac{\partial F}{\partial p_{ij}} \frac{\partial p_{ij}}{\partial \delta x} \qquad (7.2)$$

over the range of pixel indices i and j.

These partial derivatives can be computed by an analogy of backpropagation formulas of Sect. 5.2.

This approach is admittedly mathematically challenging. All desired invariances have to be precisely formulated in the form of parametric functions. This may be the reason for relatively scarce use of this concept.

The most popular is an approach that is easier to use, at the expense of larger computing costs. It is based on a simple extension of the training set. For every training example, or a selected subset thereof, several variants representing the desired invariances are generated. They may consist of (random) cropping, shifting, rotating, or mirroring. Minimizing the mapping error for such modified training examples favors mappings with the desired invariance. This simple, but frequently successful approach is sometimes referred to as *training set augmentation*.

7.3   Application of Transformer Architecture to Computer Vision

In Chap. 6 dedicated to NLP, a successful architecture called *Transformers* has been presented. This architecture has been originally developed for machine translation and then other language processing tasks but has turned out to be favorable also for some other application domains. One of them is CV. Although the architecture remains the same, there are differences in some interpretation issues that are worth mentioning. These are the topics of the following paragraphs.

7.3.1   Attention Mechanism for Computer Vision

Natural language has been presented as a sequential dynamical process extending over time (spoken language) or one-dimensional space (written language). The semantics of every word or phrase typically depends on the preceding text segments. It is assumed that future text segments cannot be processed in spoken language but are available, and can be exploited, in written text.

For longer texts, there are many candidates for delivering additional semantic information concerning the analyzed word. The most of these candidates are not relevant. The key challenge is to reduce this large candidate set to those with high relevance. This is the goal of the attention mechanism of the Transformer architecture (Sect. 6.7). It assigns weights to individual words according to their assessed relevance for the semantics of the word currently processed.

The attention principle is not limited to language processing. In fact, it is essential for intelligent processing of every voluminous sensory input. Digital images are no exception to this. For example, an image containing a cat may be disambiguated to a wildcat if the environment is a forest or a pet cat if the environment is living-room furniture. Analyzing a traffic scene for potential safety risks, a segment depicting a child may suggest increased risk if a ball can be identified in the vicinity. A flat display is probable to be TV-set in the vicinity of an audio device but be rather a computer monitor in the neighborhood of paper files.

This motivates the use of attention mechanism for CV. It can be beneficially used at the intermediary processing stage at which image segments are assigned emerging interpretation (i.e., after low-level operations such as edge extraction or trained extractions from convolutional layers have been accomplished). The corresponding neural network layer corresponds to a two-dimensional map of such emerging segments. Every one of them can be seen as an analogy to a word in a text. On such segments, similarity assessment can be done, resulting in weights expressing the relevance to the interpretation of the segment currently investigated. This is an analogy to the semantic relevance for words, as treated by the attention mechanism. The image segments used for this goals are usually *image patches* discussed in Sect. 7.3.2.

7.3.2　Division into Patches

Digital images are represented by a two-dimensional matrix. Every element of this matrix p_{ij} is typically strictly in the range of 0 to 1 indicating the whiteness of this pixel. If p_{ij} is 0, the pixel will be interpreted as *black*, for $p_{ij} = 1$, it will be interpreted as *white*. The intermediary values are scaled linearly from black to white. For colored images, the each pixel is represented by three numbers: red, green, and blue (RGB). Different color maps and interpretations exist such as HSV or CMYK; however, the overall argumentation and architecture are not influenced by the interpretation of the pixel values. For simplicity, in this chapter (as in Sect. 7.1.1), we assume a square grayscale image.

In the first appearance of the (today widespread) application of the Transformer architecture to CV (the *Vision Transformer* from [1]) it became apparent that using each pixel individually as the input is not appropriate for the attention mechanism. In NLP, the attention mechanism attempts to determine the relevance of various tokens (e.g., words in a text segment) for a given token. If a pixel was chosen to play the role of a token, it would mean that the relevance is sought between the pixels. But it is obvious that individual pixels are units of a too low level for relevance assessment. Additionally, assessment for all pairs of pixels would be prohibitively expensive. Therefore, it is necessary to select larger tokens that can realistically carry semantic information in the image.

Analogically to combining characters to form a word, a region of multiple pixels is used to form a *patch*. The input image is sliced into 16×16 patches[1], linearly

---

[1] This is valid for medium-sized input images. To maintain scalability in memory for increasing image size or to maintain training performance in small-size input images, this size must be adapted.

projected into an embedding space (the same way as words), and then treated like any other input pattern in the Transformer architecture.

The values of individual pixels within a patch are ordered into a vector. By this, positional information about the pixels involved is lost. This may seem a significant loss, but in NLP, the same logic is applied: The order of the actual characters building the word is neglected, and only the instance of the word is used.

For NLP, it is obvious that the positions of words—absolute in the sentence or paragraph or relative to another word—are relevant for semantics. Words closer to the word currently analyzed are more probable to constitute the relevant context and thus to contribute to the semantic representation. Beyond this, natural languages possess more or less strict grammars in which orders and positions result from grammar rules.

This is why positional encodings, called also position embeddings, are a key element of the semantic processing.

In CV, the role of positions may seem less strong. One reason for this is the absence of composition rules (analog to a language grammar) in an image. On the other hand, objects can be composed of parts (some of which can be occluded) that give an unambiguous interpretation to the picture. For these reasons, position embeddings analogical to that of natural language are frequently used in Transformer-based architectures for CV.

The analogy is complete for absolute positioning where the embedding is learned from the training set. For relative positions of two objects or segments, the concepts are generalized to two dimensions. For every pair of segments, their relative position is to be characterized.

This can be done in a few ways:

- by a linear numbering of the patches;
- by a pair of absolute coordinates of the patches;
- by a pair of relative distances along x and y axes;
- by the Euclidean distance of both segments.

As for one-dimensional relative position in a word sequence, these distances can be used to extend or modify the similarity weight of attention mechanism.

Once this preprocessing is finished, the further processing behaves exactly like any other (NLP-based) Transformer architecture presented in Sect. 6.5.

7.4 Section Summary

CV basically consists in the transformation of low-level signals (individual image pixels with gray intensity or color information) into the description what the whole image represents. Correspondingly, the processing is a sequence of low- to high-level operations.

The low-level operations extract simple features extending over small feature regions such as edges, contrast, or color changes. The next level implements important invariances with regard to shift, rotation, and scale. The final processing assigns the image features to labeled objects.

In neural networks, low-level operations are successfully implemented as convolution layers, corresponding to local mathematical operators such as that of Laplace. Handling invariances is a large challenge. Various approaches have been proposed, from analytical penalty terms for missing invariance to varying positions and scales in the training data.

Analogically to Natural Language Processing, image processing has to consider context aspects. This is why the attention mechanism and Transformer architecture have recently also been used for Computer Vision tasks. The role of language tokens is played by image patches (smaller regions of an image). The interpretation of image patches is modified by context formed by other patches.

7.5   Comprehension Check

1. Which low-level operations are important in Computer Vision?
2. Which are the means for implementing low-level operations in neural networks?
3. What is the effect of convolutional layers to the number of network parameters?
4. Which invariances are important for Computer Vision?
5. What are the means for implementing the invariances?
6. What is the analogy to a natural language token (or word) in an image, if exploited by the attention mechanism?
7. What is the analogy to a token position in natural language in an image?

References

1. Dosovitskiy A, Beyer L, Kolesnikov A, Weissenborn D, Zhai X, Unterthiner T, Dehghani M, Minderer M, Heigold G, Gelly S, Uszkoreit J, Houlsby N (2021) An image is worth 16×16 words: Transformers for image recognition at scale. In: International conference on learning representations, Vienna, Austria, p 21. https://openreview.net/forum?id=YicbFdNTTy
2. He K, Zhang X, Ren S, Sun J (2016) Deep residual learning for image recognition. In: 2016 IEEE conference on computer vision and pattern recognition (CVPR), pp 770–778. https://doi.org/10.1109/CVPR.2016.90
3. He K, Zhang X, Ren S, Sun J (2016) Identity mappings in deep residual networks. In: Leibe B, Matas J, Sebe N, Welling M (eds) Computer vision—ECCV 2016. Springer International Publishing, Cham, Lecture notes in computer science, pp 630–645. https://doi.org/10.1007/978-3-319-46493-0_38

4. Krizhevsky A, Sutskever I, Hinton GE (2012) ImageNet classification with deep convolutional neural networks. In: Advances in neural information processing systems. Curran Associates, Inc., vol 25. https://proceedings.neurips.cc/paper/2012/hash/c399862d3b9d6b76c8436e924a68c45b-Abstract.html
5. Lecun Y, Bottou L, Bengio Y, Haffner P (1998) Gradient-based learning applied to document recognition. Proc IEEE 86(11):2278–2324. https://doi.org/10.1109/5.726791
6. Russakovsky O, Deng J, Su H, Krause J, Satheesh S, Ma S, Huang Z, Karpathy A, Khosla A, Bernstein M, Berg AC, Fei-Fei L (2015) ImageNet large scale visual recognition challenge. Int J Comput Vis 115(3):211–252. https://doi.org/10.1007/s11263-015-0816-y
7. Simard P, Victorri B, Lecun Y, Denker JS (1991) Tangent prop - A formalism for specifying selected invariances in adaptive networks. In: Moody JM, Hanson SJ, Lippman RP (eds) NIPS'91: Proceedings of the 4th International Conference on Neural Information Processing Systems, Morgan Kaufmann, pp 895–903. https://doi.org/10.5555/2986916.2987026
8. Simonyan K, Zisserman A (2015) Very deep convolutional networks for large-scale image recognition. https://arxiv.org/abs/1409.1556
9. Wechsler H (1990) Computational vision, Computer science and scientific computing. Academic Press, Boston

Index

A

Activation function, 12, 56, 60, 136
  logistic, *see* sigmoid
  piecewise linear, 61
  ReLU, 57, 93, 184, 189, 198
  sigmoid, 11, 57, 61, 65, 93, 126, 176, 189
    symmetric, 64
  softmax, 178
  tanh, 57, 176
  threshold, 57, 61
AdaGrad, 145
Adaline, 144
Adam, 145
Adaptive step size, 144
Artificial Intelligence (AI), 1, 2, 83, 145
Attention mechanism, 177–180, 192
  attention heads, 183
  key vector, 179, 182
  multi-head, 188
  multi-head self-attention, 190
  position-wise feedforward, 184
  query vector, 179, 182
  self-attention, 182, 184, 188
  value vector, 179, 182
Autoregressive Moving Average (ARMA), 37

B

Backpropagation, 136, 189, 203
Backward pass, 137
Bag-of-word model, 172
Bayesian, 32, 59, 115, 130
  inference, 113
  network, 58
  optimum, 22, 25, 32, 114

  point estimate, 114
  posterior, 21
  regularization, 109
Bidirectional Encoder Representations from
    Transformers (BERT), 48, 143, 177, 180,
    181
Bode plot, 153, 155
Boltzmann
  constant, 158
  distribution, 158
  statistics, 158
Bootstrap, 119

C

Classification, 1, 2, 8, 15, 21, 42, 49, 130
  image, 41, 42, 195, 197
  text, 42, 169, 172
Computer Vision (CV), 3, 195, 200, 203–205
Conjugate, 138
  gradient, 139–141, 143, 144
  projection, 139
Contextualized representation, 181
Context vector, 178, 182, 183
Convolutional layer, 200, 202
Convolutional Neural Network (CNN), 197,
    201
Convolutional operator, 196, 197
  filter, 197
  kernel, 197
Corpus-based semantics, 167
Cross-validation, 118, 120

D

Data

Printed in the United States
by Baker & Taylor Publisher Services